Springer Undergraduate Mathematics Series

Advisory Board

Other books in this series

A First Course in Discrete Mathematics *I. Anderson*
Analytic Methods for Partial Differential Equations *G. Evans, J. Blackledge, P. Yardley*
Applied Geometry for Computer Graphics and CAD, Second Edition *D. Marsh*
Basic Linear Algebra, Second Edition *T.S. Blyth and E.F. Robertson*
Basic Stochastic Processes *Z. Brzeźniak and T. Zastawniak*
Calculus of One Variable *K.E. Hirst*
Complex Analysis *J.M. Howie*
Elementary Differential Geometry *A. Pressley*
Elementary Number Theory *G.A. Jones and J.M. Jones*
Elements of Abstract Analysis *M. Ó Searcóid*
Elements of Logic via Numbers and Sets *D.L. Johnson*
Essential Mathematical Biology *N.F. Britton*
Essential Topology *M.D. Crossley*
Fields and Galois Theory *J.M. Howie*
Fields, Flows and Waves: An Introduction to Continuum Models *D.F. Parker*
Further Linear Algebra *T.S. Blyth and E.F. Robertson*
Geometry *R. Fenn*
Groups, Rings and Fields *D.A.R. Wallace*
Hyperbolic Geometry, Second Edition *J.W. Anderson*
Information and Coding Theory *G.A. Jones and J.M. Jones*
Introduction to Laplace Transforms and Fourier Series *P.P.G. Dyke*
Introduction to Lie Algebras *K. Erdmann and M.J. Wildon*
Introduction to Ring Theory *P.M. Cohn*
Introductory Mathematics: Algebra and Analysis *G. Smith*
Linear Functional Analysis *B.P. Rynne and M.A. Youngson*
Mathematics for Finance: An Introduction to Financial Engineering *M. Capiński and T. Zastawniak*
Matrix Groups: An Introduction to Lie Group Theory *A. Baker*
Measure, Integral and Probability, Second Edition *M. Capiński and E. Kopp*
Metric spaces *M.Ó. Searcóid*
Multivariate Calculus and Geometry, Second Edition *S. Dineen*
Numerical Methods for Partial Differential Equations *G. Evans, J. Blackledge, P.Yardley*
Probability Models *J. Haigh*
Real Analysis *J.M. Howie*
Sets, Logic and Categories *P. Cameron*
Special Relativity *N.M.J. Woodhouse*
Symmetries *D.L. Johnson*
Topics in Group Theory *G. Smith and O. Tabachnikova*
Vector Calculus *P.C. Matthews*

N.M.J. Woodhouse

General Relativity

With 33 Figures

 Springer

N.M.J. Woodhouse
Mathematical Institute
24-29 St Giles'
Oxford OX1 3LB
UK

Cover illustration elements reproduced by kind permission of:
Aptech Systems, Inc., Publishers of the GAUSS Mathematical and Statistical System, 23804 S.E. Kent-Kangley Road, Maple Valley, WA 98038, USA. Tel: (206) 432 -7855 Fax (206) 432 -7832 email: info@aptech.com URL: www.aptech.com.
American Statistical Association: Chance Vol 8 No 1, 1995 article by KS and KW Heiner 'Tree Rings of the Northern Shawangunks' page 32 fig 2.
Springer-Verlag: Mathematica in Education and Research Vol 4 Issue 3 1995 article by Roman E Maeder, Beatrice Amrhein and Oliver Gloor 'Illustrated Mathematics: Visualization of Mathematical Objects' page 9 fig 11, originally published as a CD ROM 'Illustrated Mathematics' by TELOS: ISBN 0-387-14222-3, German edition by Birkhauser: ISBN 3-7643-5100-4.
Mathematica in Education and Research Vol 4 Issue 3 1995 article by Richard J Gaylord and Kazume Nishidate 'Traffic Engineering with Cellular Automata' page 35 fig 2. Mathematica in Education and Research Vol 5 Issue 2 1996 article by Michael Trott 'The Implicitization of a Trefoil Knot' page 14.
Mathematica in Education and Research Vol 5 Issue 2 1996 article by Lee de Cola 'Coins, Trees, Bars and Bells: Simulation of the Binomial Process' page 19 fig 3. Mathematica in Education and Research Vol 5 Issue 2 1996 article by Richard Gaylord and Kazume Nishidate 'Contagious Spreading' page 33 fig 1. Mathematica in Education and Research Vol 5 Issue 2 1996 article by Joe Buhler and Stan Wagon 'Secrets of the Madelung Constant' page 50 fig 1.

Mathematics Subject Classification (2000): 83-01

British Library Cataloguing in Publication Data
A catalogue record for this book is available from the British Library

Library of Congress Control Number: 2006926445

Springer Undergraduate Mathematics Series ISSN 1615-2085
ISBN-10: 1-84628-486-4   e-ISBN 1-84628-487-2   Printed on acid-free paper
ISBN-13: 978-1-84628-486-1

9 8 7 6 5 4 3 2 1

Springer Science+Business Media, LLC
springer.com

Preface

It is a challenging but rewarding task to teach general relativity to undergraduates. Time and experience are in short supply. One can rely neither on the undivided attention of students who are studying many other exciting topics in the final years of their course, nor on easy familiarity with the classical tools of applied mathematics and geometry. Not only are the ideas themselves difficult, but the calculations needed to solve even quite simple problems are themselves technically challenging for students who have only recently learned about multivariable calculus and partial differential equations.

For those with a strong background in pure mathematics, there is the temptation to present the theory as an application of differential geometry without conveying a clear understanding of its detailed connection with physical observation. At the other extreme, one can focus too exclusively on physical prediction, and ask the audience to take too much of the mathematical argument on trust.

This book is based on a course given at the Mathematical Institute in Oxford over many years to final-year mathematics students. It is in the tradition of physical applied mathematics as it is taught in this country, and may, I hope, be of use elsewhere. It is coloured by the mathematical leaning of our students, but does not present general relativity as a branch of differential geometry. The geometric ideas, which are of course central to the understanding of the nature of gravity, are introduced in parallel with the development of the theory—the emphasis being on laying bare how one is led to pseudo-Riemannian geometry through a natural process of reconciliation of special relativity with the equivalence principle. At centre stage are the 'local inertial coordinates' set up by an observer in free-fall, in which special relativity is valid over short times and distances.

In more practical terms, the book is a sequel, with some overlap in the treatment of tensors, to my *Special Relativity* in this same series. The first nine chapters cover the material in the Mathematical Institute's introductory lectures. Some of the material in the last three chapters is contained in a second set of lectures that has a more fluid syllabus; the rest I have added to introduce the theoretical background to contemporary observational tests, in particular the detection of gravitational waves and the verification of the Lens–Thirring precession. I have also added some sections (marked *) which can be skipped.

There are a number of very good books on relativity, some classic and some more recent. I hope that this will be a useful if modest addition to the collection. I have drawn in particular on the excellent books by Misner, Thorne and Wheeler [14], Wald [22], and Hughston and Tod [9]. I also acknowledge the help of my colleagues who have shared the teaching of relativity in Oxford over the years, particularly Andrew Hodges, Lionel Mason, Roger Penrose, and Paul Tod. Most of the problems in the book are ones that have been used by us many times on problem sheets, and their origin is sometimes forgotten. Inasmuch as they may originally have been adapted from other texts, I apologise for being unable to cite the original sources. I am grateful for the hospitality of the Isaac Newton Institute in Cambridge in September 2005. Part of this book was written there during the programme *Global problems in mathematical relativity*.

Oxford, February 2006   NMJW

Contents

1
Newtonian Gravity

1.1 'Special' and 'General' Relativity

Even before Newton had written down the laws of motion, Galileo had observed that it is impossible to detect uniform motion in an enclosed space. If you do experiments in the cabin of a ship on a calm sea—for example, by dripping water into a bucket or by observing the flight of insects—then you will get the same results whether the ship is moving uniformly or at rest. The common motion of the ship and the objects of the experiment has no detectable effect.

The observation has a precise formulation within the framework of classical dynamics, in the statement that the laws of motion are invariant under Galilean transformations. Start with a frame of reference in which Newton's laws are valid, and use Cartesian coordinates x, y, z to measure the positions, velocities, and accelerations of moving bodies. Then the assertion is that they remain valid when we replace x, y, z by the Cartesian coordinates x', y', z' of a new frame of reference in uniform motion relative to the original one. Two such coordinate systems are related by a *Galilean transformation*

$$\begin{pmatrix} x \\ y \\ z \end{pmatrix} = H \begin{pmatrix} x' \\ y' \\ z' \end{pmatrix} + \begin{pmatrix} a + ut \\ b + vt \\ c + wt \end{pmatrix}, \tag{1.1}$$

where H is a constant rotation matrix, t is time, and a, b, c, u, v, w are constants. The constancy of H implies that the new frame is not rotating relative to the old; but its origin moves with constant velocity (u, v, w) relative to the old frame.

Put another way, there is no absolute standard of rest in classical mechanics. Instead there is a special class of frames of reference, called *inertial frames*, in which the laws of motion hold. The coordinate systems of any two inertial frames are related by a Galilean transformation. No inertial frame is picked out as having the special status of being at rest, but any two are in uniform motion relative to each other. This is encapsulated in the *principle of relativity*, that in classical mechanics all inertial frames are on an equal footing. No mechanical experiment will detect absolute motion: only relative motion has physical meaning.

Maxwell's equations, on the other hand, are not invariant under Galilean transformations. They appear to single out a particular set of frames as being 'at rest': that is, to imply that it should be possible to detect absolute motion by electromagnetic experiments. This could, of course, be 'motion relative to the ether', the all-pervasive but undetected medium that was supposed to propagate electromagnetic waves in the original nineteenth century theory. But Einstein arrived at a more satisfactory resolution of the unwelcome violation of relativity without appeal to this fictitious substance: that the principle of relativity does extend to electromagnetism, but that the transformation between inertial frames is not a Galilean transformation. It is instead the *Lorentz transformation*. I assume that the reader is already familiar with this story and do not repeat it here.

In both the classical world and in Einstein's special theory of relativity, inertial frames are characterized by the absence of acceleration and rotation. Acceleration and angular velocity are absolute. An observer can tell whether a frame of reference is inertial without reference to any other frame, by seeing whether Newton's first law holds. If particles that are not subject to a force move relative to the frame in straight lines at constant speed, then the frame is inertial; if they do not, then it is not. More simply, rotation and acceleration can be 'felt'.

The situation is less clearcut when gravity enters the picture. Because the gravitational and inertial masses of a body are the same, it is impossible to tell the difference, locally, between the effects of acceleration of the frame of reference and those of gravity. An observer who falls towards the laboratory floor may be seeing the effects of gravity, or simply, but perhaps less plausibly, the effects of the acceleration of the laboratory in the upward direction. No local experiment within the laboratory will distinguish the two possibilities. In Newtonian gravity, the distinction is a global one: in a nonaccelerating frame, the apparent gravitational field vanishes at large distances; in an accelerating frame it takes a nonzero constant value at infinity.

I expand on these remarks below, after a brief review of Newtonian gravitation. But the broad conclusion is already clear: a theory of gravity must address the transformation between accelerating frames. Special relativity deals only with 'special' coordinate transformations between the coordinates of inertial frames. Gravity requires us to look at 'general' transformations between frames in arbitrary relative motion.

1.2 Newton's Theory

The essential content of Newton's theory of gravity is contained in two equations. The first is Poisson's equation

$$\nabla^2 \phi = 4\pi G \rho, \tag{1.2}$$

where ϕ is the gravitational potential, ρ is the matter density, and G is the *gravitational constant*, with dimensions $L^3 M^{-1} T^{-2}$ and value 6.67×10^{-11} in SI units. With appropriate boundary conditions, it determines the gravitational potential of a given source. The second equation relates the gravitational field to the gravitational potential, by

$$\boldsymbol{g} = -\boldsymbol{\nabla}\phi. \tag{1.3}$$

It determines the force $M\boldsymbol{g}$ on a particle of mass M. If the particle is falling freely with no other forces acting, then the total energy

$$E = \tfrac{1}{2}Mv^2 + M\phi$$

is constant during the motion. It is sum of the kinetic energy $\frac{1}{2}Mv^2$, where v is the speed, and the *potential energy $M\phi$*. Hence the term 'potential'.

The accuracy of the theory is remarkable: in the solar system, the only detectable discrepancy between the theoretical and actual motions of the planets is in the orbit of Mercury, where it amounts to one part in 10^7.

The two equations contain the inverse square law. By integrating over a region bounded by a surface S, and containing a total mass m, we obtain Gauss's law from the divergence theorem:

$$\int_S \boldsymbol{g}.\mathrm{d}\boldsymbol{S} = -4\pi G m. \tag{1.4}$$

If the field is spherically symmetric, for example, if it is that outside a spherical star, then the magnitude g of $\boldsymbol{g}$ depends only on the distance r from the centre of the star and the direction of $\boldsymbol{g}$ is towards the centre. By taking S to be a sphere of radius r, we obtain

$$g = \frac{Gm}{r^2},$$

which is the inverse square law. The corresponding potential is not unique because we are free to add a constant. If we fix this by taking $\phi = 0$ at infinity, then $\phi = -Gm/r$ and the energy of a particle of mass M falling under the influence of the star's gravity is

$$E = \frac{Mv^2}{2} - \frac{GMm}{r}\,. \tag{1.5}$$

1.3 Gravity and Relativity

Newton's theory of gravity is consistent with Galilean relativity. If the equations (1.2) and (1.3) hold in one inertial frame of reference, then they hold in every inertial frame. In particular, if we transform from one frame to a second in uniform motion by replacing the Cartesian coordinates $x, y,$ and z in the first frame by $x', y',$ and z', where

$$x = x', \qquad y = y', \qquad z = z' + vt\,, \tag{1.6}$$

then the accelerations of particles are the same in the new frame and in the old, and

$$\boldsymbol{\nabla}\phi = \boldsymbol{\nabla}'\phi\,,$$

where $\boldsymbol{\nabla}'$ is the gradient in the new coordinates. So Poisson's equation and the relationship between the gravitational acceleration and the gradient of the potential are still valid in the second frame.

Thus far there is no problem. The theory can be tested against observation by using Poisson's equation to predict the gravitational field $\boldsymbol{g}$ of a given distribution of matter, and then by verifying that the motion of a particle is governed by the equation

$$M\ddot{\boldsymbol{r}} = M\boldsymbol{g}\,, \tag{1.7}$$

where $\boldsymbol{r}$ is its position vector from the origin. We see here, however, the first hint of difficulty, in the equality of the two Ms on the left- and right-hand sides. Their cancellation has a consequence which at first sight seems merely convenient, but which on deeper thought raises a question about the physical identity of the gravitational field. The implication is that Newton's theory is also invariant under another type of transformation, a *uniform acceleration*. If instead of (1.6), we put

$$x = x', \qquad y = y', \qquad z = z' + \tfrac{1}{2}at^2\,, \tag{1.8}$$

where a is a constant acceleration, then the accelerations $\ddot{\boldsymbol{r}}$ and $\ddot{\boldsymbol{r}}'$ of a particle in the two coordinate systems are related by

$$\ddot{\boldsymbol{r}} = \ddot{\boldsymbol{r}}' + \boldsymbol{a}\,,$$

where $\boldsymbol{u}$ is a vector of magnitude u in the z direction. Then (1.7) holds in the new coordinates provided we replace $\boldsymbol{g}$ by $\boldsymbol{g}' = \boldsymbol{g} - \boldsymbol{a}$, because we then have

$$M\ddot{\boldsymbol{r}}' = M(\ddot{\boldsymbol{r}} - \boldsymbol{a}) = M(\boldsymbol{g} - \boldsymbol{a}) = M\boldsymbol{g}'\,.$$

Equivalently, if we replace ϕ by $\phi' = \phi - az$ when we transform to the accelerating coordinate system, then (1.3) still holds; and because the second derivatives of ϕ and ϕ' with respect to the Cartesian coordinates are unchanged, Poisson's equation also holds in the new system. Of course there is nothing special about the z-direction: we draw a similar conclusion whatever the direction of the acceleration $\boldsymbol{a}$.

Thus Newton's theory of gravity also holds in any uniformly accelerating frame of reference, provided that we subtract the acceleration $\boldsymbol{a}$ from $\boldsymbol{g}$ when we transform from one frame of reference to another accelerating relative to it with acceleration $\boldsymbol{a}$. In particular, we can make the gravitational field at a point appear to vanish by taking $\boldsymbol{a}$ to be the value of $\boldsymbol{g}$ at the point: then $\boldsymbol{g}' = 0$ in the accelerating coordinate system. That is, gravity is unobservable at a point in a frame in *free-fall*, in which massive particles will appear to be 'weightless'.

The phenomenon is more familiar now than in Newton's day. Astronauts, for example, are trained to cope with weightlessness by flying them in an aeroplane accelerating towards the earth with the acceleration due to gravity; and of course the weightlessness they experience in space travel is not, as is often incorrectly reported, because they are 'beyond the earth's gravity', but because, with rocket motors not firing, their spacecraft is in free-fall—they are falling with acceleration equal to the local gravitational field $\boldsymbol{g}$.

So how can one disentangle the 'true' gravitational field from the 'apparent' one derived from the acceleration of the frame in which measurements are made? Locally, one cannot: the nonaccelerating frames of reference are distinguished from the accelerating ones only by the fact that in a nonaccelerating frame, the gravitational field falls to zero a long way from the source. The distinction is a global one.

In fact the gravitational field that we measure on the surface of the earth is a combination of the 'true' field—generated principally by the attraction of the earth itself—and the effects of acceleration due to the rotation of the earth and to its orbital motion around the sun. The true field—the field in an inertial frame—has to be calculated by correcting the *apparent gravity* for the effects of acceleration. It is apparent gravity that is measured by weighing an object of unit mass at rest on the earth's surface.

In the classical theory, the distinction between real fields and apparent ones is clear, and the inclusion of the acceleration of the frame in the apparent gravitational field is seen as simply a computational device to deal with problems in which it is convenient to work in an accelerating frame instead of an inertial one.

It is when we try to include gravitation in special relativity that the issue of the reality of the distinction comes to the fore. How is an observer in a gravitational field to identify the inertial frames of the special theory of relativity? They are supposed to be the frames in which Newton's first law holds:

> In the absence of forces, particles move in straight lines at constant speed.

The difficulty is that gravity affects all matter equally, so there are no completely free particles. It also affects light, as we show shortly. So it is not possible simply to adapt the classical definition, that the frame of an observer in a gravitational field is inertial if it is not accelerating relative to a distant inertial observer a long way from the source. The two observers would need to exchange light signals to measure their relative acceleration, and these would be affected by gravity.

This problem did not arise in electromagnetic theory because there are charged particles that are affected by an electric field and other neutral particles that are not. The motion of neutral particles can, in principle, be used to pick out the inertial frames, and the fields can then be determined from the behaviour of charged particles; but in gravitation theory, there are no 'neutral' particles which we can think of as free of all forces.

1.4 The Equivalence Principle

It is impossible for an observer to distinguish the local effects of gravity and acceleration only because the Ms on the two sides of the equation of motion (1.7) cancel. Before we go further, we should pause and ask if this is really true. Is the cancellation exact for all forms of matter? The answer will have a profound influence on our view of the physical nature of gravity.

The M on the left-hand side of (1.7) is the *inertial mass*, which determines the way in which a body reacts to force in Newton's second law; that on the right-hand side is the *gravitational mass*. It is analogous to charge in electromagnetism: it determines the force experienced by a body in a given field. If they were not always equal, then the acceleration due to gravity would not be the same for all types of matter.

Their equality was first tested by Galileo, by comparing the periods of pendula with weights made out of different materials. He found no difference. A more precise confirmation came from the celebrated nineteenth century experiment by Eötvös, which verified that the two types of mass are equal at least to one part in 10^9 [2]. His idea was that if the two masses were not always the same, then the apparent gravitational field at the earth's surface should depend on the composition of a body. If the equality failed, then it should be possible to find two bodies with equal gravitational mass, but unequal inertial mass. The attraction of the earth would be the same for both, but the acceleration corrections would be different. The latter have horizontal components. So if the two bodies were fixed to opposite ends of a rod suspended at its centre by a thin wire, then the rod should twist. When the masses are interchanged, it should twist other way. The effect was likely to be very small, and the experiment very delicate, but Eötvös found no evidence for any difference in the two types of mass. More recent experiments, including lunar ranging measurements, have reinforced the conclusion at the level of one part in 10^{13}; and a planned space experiment STEP will test it to one part in 10^{18} [20].

We have good reason, therefore, to accept the (weak) *equivalence principle*, which is that the equality is exact and intrinsic to the nature of gravity. Einstein went further, and based general relativity on the assumed truth of the *strong equivalence principle*.

> There is no observable distinction between the local effects of gravity and acceleration.

There is no physical experiment that can be performed within an isolated room that will reveal whether (i) the room is at rest on the earth's surface, or (ii) it is in a spaceship accelerating at the acceleration due to gravity in the direction of the ceiling in otherwise empty space. In both cases, those inside the room 'feel' a normal terrestrial gravitational field. The principle asserts that all experiments that do not involve looking at the outside environment will similarly fail to distinguish the two situations.

1.5 Linearity and Light

The equivalence principle poses a challenge to any attempt to incorporate gravitational fields within the framework of special relativity, in the same way as electromagnetic theory. It undermines the identification of inertial frames: if the effects of gravity and acceleration are locally indistinguishable, how do you pick out the nonaccelerating frames of special relativity?

It also raises a more subtle problem. Maxwell's equations and Poisson's equation are linear. If you superimpose two charge distributions or two mass distributions then the electromagnetic or gravitational field of the combined distribution is simply the sum of the individual fields. However, as bodies interact gravitationally, energy is transferred from their gravitational fields to the bodies themselves, and vice versa. Thus gravitational fields themselves carry energy, and therefore have inertial mass, a consequence of the fundamental relativistic equality of mass and energy. But if inertial mass and gravitational mass are the same, then gravitational fields must themselves generate gravitational fields. If two large masses are brought close together, then the potential energy of one in the field of the other must be accounted for in the total energy of the combined system, and must contribute to the total gravitational field. Simple addition of the individual fields will not allow this. Thus a relativistic theory of gravity must be based on nonlinear equations. It cannot be founded on Lorentz-invariant linear equations that look anything like Maxwell's equations.

A similar problem dogs any naive attempt to combine classical gravitational theory with electrodynamics. One aspect of this can be seen in the energy conservation equation (1.5). For a particle in a gravitational field of the spherical star to escape to infinity, its speed v must exceed the escape velocity

$$v = \sqrt{2Gm/r}$$

because E is conserved and v^2 must remain nonnegative as $r \to \infty$. The escape velocity is maximal at the star's surface, where r takes its lower possible value in the region outside the star. What if at this point we have $v = c$, the velocity of light? This will be the case if the radius of the star is $R = 2Gm/c^2$, the so-called *Schwarzschild radius*. Then nothing can escape from the surface. But what if there is a mirror on the surface and we shine light down from infinity? It will be reflected at the surface and follow the same path back out again. Because orbits are reversible in Newtonian gravity, it will be reflected back to infinity. Clearly Newtonian theory does not provide a consistent picture of such a 'black hole', because it does not allow for a consistent picture of the interaction of light and gravity. Photons carry energy, and so must be affected by gravity, something that is at odds with the notion of a universally constant 'speed of light'. This is intertwined with the previous problem: the constant 'speed of light' in special relativity is the speed of light relative to an inertial frame. We cannot even say what the 'speed of light' means in the presence of gravity without first identifying the inertial frames.

The argument that 'photons carry energy and must therefore be affected by gravity' is made more fully by Bondi's *gedanken* experiment: he showed that if photons were not affected by gravity, then one could in principle build a perpetual motion machine. He imagined a machine consisting of a series of buckets attached to a conveyor belt. Each contains a single atom, with those on the right in an excited state and those on the left in a lower energy state. As they reach the bottom of the belt, the excited atoms emit light which is focused by two curved mirrors onto the atom at the top of the belt, the one at the bottom falls into the lower state and the one at the top is excited. Because $E = mc^2$, those on the right, which have more energy, should be heavier. The force of gravity should therefore keep the belt rotating in perpetuity.

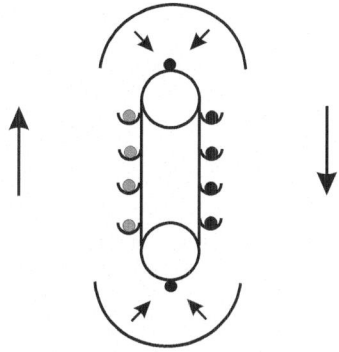

Figure 1.1 Bondi's *perpetuum mobile*

The resolution is that photons lose energy as they climb up through the gravitational field. Because $E = \hbar\omega$, they must therefore be redshifted. This was confirmed directly by Pound and Rebka in 1959 in a remarkable experiment in which they measured the shift over the 75 ft height of the tower of the Jefferson building at Harvard [16]. It is about 3 parts in 10^{14}.

Pound and Rebka's result is incompatible with special relativity, as can be seen from the space–time diagram, Figure 1.2. The vertical lines are the histories of the top and bottom of the tower and the dashed lines at 45° are the worldlines of photons travelling up the tower. Because the top and bottom of the tower are at rest relative to each other, their worldlines in special relativity are parallel, which forces $\Delta t = \Delta t'$. So in a special-relativistic theory of gravity, there cannot be any gravitational redshift.

The interaction of light and gravity has been observed directly, first and most famously in Eddington's observations during the 1919 eclipse of the sun, and more recently and dramatically in the pictures taken by the Hubble space telescope of distorted images of distant galaxies produced by 'gravitational lensing'. Eddington confirmed that the path followed by light reaching the earth from a star in the direction of the sun is bent by the sun's gravitational field. During a total eclipse, one can see the star field in the direction of the sun and compare the apparent positions of stars in the sky with pictures taken at night at another time of year when the sun is in a different part of the sky.

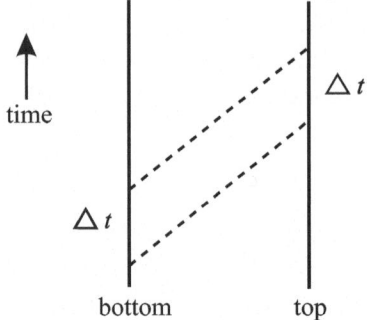

Figure 1.2 Pound and Rebka's measurement is incompatible with special relativity

Eddington observed that the apparent positions of the stars close to the edge of the sun were displaced outwards as the light rays from them were bent inwards as they passed the sun.[1]

1.6 The Starting Point

In a gravitational field, it is impossible to identify the global inertial frames of special relativity by local observation. We can, however, pick out local inertial frames in which gravity is 'turned off': a local inertial frame is one set up by an observer in free-fall, by using a clock and light signals to assign coordinates to nearby events. It follows from the equivalence principle that, provided that the observer makes observations only in a small neighbourhood of a given event on his worldline, then the usual framework of nongravitational physical theory should hold good, and the transformation between local inertial coordinate systems in the neighbourhood will be the same as in special relativity, at least as an approximation over short times and small distances.

But if we can only work in frames in which gravity is turned off, then how can we observe gravity? The answer is, by a shift in point of view. Gravity is not seen in the 'force' exerted on a massive body, but rather in the relative acceleration of nearby local inertial observers. If they make measurements only over short distances and times, then two nearby observers in spaceships in

---

[1] The history of Eddington's observation is not quite as straightforward as it is sometimes presented. See, for example, Peter Coles' article [5].

free fall towards the earth's gravitational field cannot tell that they are in a gravitational field and not simply accelerating uniformly in empty space a long way from any source of gravitation. If, however, they are farther apart, then there will be a small relative acceleration between them because the earth's field is not uniform. From the point of view of someone standing on the earth's surface, the relative acceleration is the difference in values of g at their two locations.

Although an observer in a spaceship in orbit cannot detect gravity by local measurements, it is not necessary to consider the distant environment to detect its presence: the observer can distinguish between real and apparent gravity by tracking the small relative acceleration of nearby objects in free-fall.

So the big step that we make to accommodate the equivalence principle is to ignore the gross effect of gravity, the 'acceleration due to gravity', which is indistinguishable from the apparent gravity in an accelerating frame, and to regard as primary the relative acceleration that it produces between nearby objects in free-fall. The physically central quantity is then not g but rather its derivatives with respect to the spatial coordinates. These are unchanged by the transformation (1.8). A mass distribution generates a nonuniform field, which varies from point to point. A uniform field has no observer-independent significance: it can be reduced to zero everywhere simultaneously by switching to an accelerating frame.

With this shift in viewpoint, we can begin to develop a theory of gravity that incorporates special relativity by taking as our starting point that special relativity should hold in frames in free-fall. But we can only require that it holds locally, in time and space, because we expect the effects of gravity to manifest themselves in small corrections to the Lorentz transformation between the inertial coordinate systems set up by nearby observers: their relative acceleration will destroy the exact linearity of the transformation.

> *Starting point.* Special relativity holds over short distances and times in frames in free-fall. Gravity is not a local force field, but shows up in the small relative acceleration between local inertial frames. In the presence of gravity, the transformation between local 'inertial' coordinates is not exactly linear.

The idea of *curvature* comes in here, by analogy with mapmaking. If one makes maps of the earth's surface by projecting onto a tangent plane from the centre of the earth, then overlapping maps will be slightly distorted relative to each other because of the curvature of the earth. To the first order, the transformation between the x, y coordinates on two overlapping maps will be linear, but the curvature of the earth prevents it from being exactly linear.

Figure 1.3 Relative acceleration in free fall

EXERCISES

1.1. By applying Gauss's theorem, derive the internal and external gravitational potentials for a solid uniform sphere, mass m, radius a.

1.2. By starting with the inverse square law $\boldsymbol{g} = -Gmr^{-3}\boldsymbol{r}$ for the gravitational field of a fixed point mass m, obtain the equations of motion of a test particle in plane polar coordinates r, θ. Show that if $u = Gm/r$ is expressed as a function of θ, then

$$\tfrac{1}{2}(p^2 + u^2) = \beta^2 u + k,$$

where $\beta = Gm/J$, $p = \mathrm{d}u/\mathrm{d}\theta$, and k and J are constants whose significance should be explained. Plot the curves traced out in the p, u-plane by the motion of the test particle for fixed k and varying values of β in the cases (i) $k > 0$, (ii) $k = 0$, and (iii) $k < 0$, and interpret them in terms of the motion of the test particle. (That is, plot the phase portraits: it may help to look at the first chapter of Jordan and Smith [10]. We repeat this exercise in general relativity. The phase portraits enable one to see at a glance how the pattern of relativistic orbits around a black hole differs from the classical case.)

1.3. A pendulum consists of a light rod and a heavy bob. Initially it is at rest in vertical stable equilibrium. The upper end is then made to accelerate down a straight line which makes an angle α with the horizontal with constant acceleration f. Show that in the subsequent motion, the pendulum oscillates between the vertical and horizontal positions if $y = f(\cos \alpha + \sin \alpha)$. (This problem is very easy if you apply the equivalence principle and think about the direction of the apparent gravitational field in an appropriate frame.)

1.4. A hollow plastic ball is held at the bottom of a bucket of water and then released. As it is released, the bucket is dropped over the edge of a cliff. What happens to the ball as the bucket falls?

1.5. A version of the following 'equivalence principle' device was constructed as a birthday present for Albert Einstein [4]. Simplified, the device consists of a hollow tube with a cup at the top, together with a metal ball and an elastic string. When the tube is held vertical, the ball can rest in the cup. The ball is attached to one end of the elastic string, which passes through a hole in the bottom of the cup, and down the hollow centre of the tube to the bottom, where its other end is secured. You hold the tube vertical, with your hand at the bottom, the cup at the top, and with the ball out of the cup, suspended on its elastic string. The tension in the string is not quite sufficient to draw the ball back into the cup. The problem is to find an *elegant* way to get the ball back into the cup.

Figure 1.4 Einstein's birthday present

2

Inertial Coordinates and Tensors

Before we take further the development of the relativistic theory of gravity, we need to establish an appropriate mathematical framework for special relativity. This must survive in the general theory as the formalism for describing local observations made by observers in free-fall in a gravitational field. In this chapter, familiarity with special relativity is assumed: the purpose is not to derive special relativity, but to introduce the language in which it will be extended to general relativity.

2.1 Lorentz Transformations

The special theory of relativity describes the relationship between physical observations made by different nonaccelerating observers, in the absence of gravity.

Each such observer labels events in space–time by four inertial coordinates t, x, y, z. At the heart of the theory is the description of the operations by which, in principle, these coordinates are measured. One does not begin, as in classical dynamics, by taking 'time' and 'distance' as having absolute and self-evident meanings derived from physical intuition; rather they are defined in terms of the operations of measuring them. The key departure from classical ideas is that the constancy of the velocity of light—its independence of direction and of the motion of the observer—is built into the definitions, so the conflict between the principle of relativity and the properties of electromagnetic waves is removed at the most fundamental level.

There are different but essentially equivalent ways of formulating the operational definitions. The one that we keep in mind is used in Bondi's k-calculus, and is based on Milne's 'radar' definition [3]. Each inertial observer carries a clock of standard design, which can be used to measure the time of events at the observer's location, and a device for measuring the direction from which light reaches the observer from a remote source. The device must not rotate, so the observer can determine whether two photons arriving at different times came from the same direction. So we note that special relativity requires that it should be possible to pick out nonaccelerating and nonrotating frames. In the absence of gravity, this is reasonable: acceleration and rotation can be 'felt'.

The observer assigns a distance and a time to a distant event E by timing the emission and and arrival times of photons. If a photon leaves the observer at time t_1, is reflected at the event E, and arrives back at time t_2, then the observer defines the time of t of E and its distance D by

$$t = \tfrac{1}{2}(t_1 + t_2), \qquad D = \tfrac{1}{2}(t_2 - t_1)$$

(see the space–time diagram, Figure 2.1). The observer can determine the di-

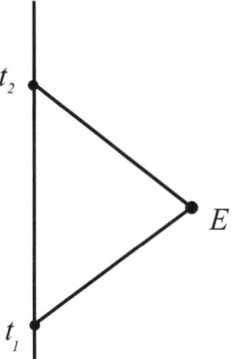

Figure 2.1 Radar definition

rection to E by observing the direction from which the returning photon arrives. Knowing the time, space, and direction of E, the observer can compute its space–time coordinates t, x, y, z. The result is an *inertial coordinate system* t, x, y, z, a term we use somewhat loosely as interchangeable with *inertial frame*. Built into the definition is the assumption that light travels with unit velocity in all directions.[1]

---

[1] We take $c = 1$ throughout.

It follows from the assumptions of special relativity that the coordinate systems t, x, y, z and $\tilde{t}, \tilde{x}, \tilde{y}, \tilde{z}$ of two inertial observers are related by an inhomogeneous Lorentz transformation

$$\begin{pmatrix} t \\ x \\ y \\ z \end{pmatrix} = L \begin{pmatrix} \tilde{t} \\ \tilde{x} \\ \tilde{y} \\ \tilde{z} \end{pmatrix} + T \,, \tag{2.1}$$

where T is a column vector, which shifts the origin of the coordinates, and

$$L = \begin{pmatrix} L^0_{\ 0} & L^0_{\ 1} & L^0_{\ 2} & L^0_{\ 3} \\ L^1_{\ 0} & L^1_{\ 1} & L^1_{\ 2} & L^1_{\ 3} \\ L^2_{\ 0} & L^2_{\ 1} & L^2_{\ 2} & L^2_{\ 3} \\ L^3_{\ 0} & L^3_{\ 1} & L^3_{\ 2} & L^3_{\ 3} \end{pmatrix} \tag{2.2}$$

is a proper orthochronous Lorentz transformation matrix.[2] This means that $L^0_{\ 0} > 0$, $\det L = 1$, and $L^t g L = g$, where

$$g = \begin{pmatrix} 1 & 0 & 0 & 0 \\ 0 & -1 & 0 & 0 \\ 0 & 0 & -1 & 0 \\ 0 & 0 & 0 & -1 \end{pmatrix} \,.$$

Each observer reckons that the other in moving in a straight line with constant speed u, given by $L^0_{\ 0} = 1/\sqrt{1 - u^2}$. The assumptions therefore exclude gravity.

Example 2.1 (Boost)

For a *boost* along the x-axis, $T = 0$ and

$$L = \begin{pmatrix} \gamma & \gamma u & 0 & 0 \\ \gamma u & \gamma & 0 & 0 \\ 0 & 0 & 1 & 0 \\ 0 & 0 & 0 & 1 \end{pmatrix} \,, \tag{2.3}$$

where $\gamma = 1/\sqrt{1 - u^2}$. In this case, the two observers have aligned their x-axes, and each is travelling along the x-axis of the other with speed u. The origin of both coordinate systems is the event at which they meet.

---

[2] The reason for departing from the standard practice of using lower indices to label the entries in a matrix will emerge shortly. The qualification 'inhomogeneous' indicates that the general transformation involves translation of the space–time coordinates. We use the term 'Lorentz transformation' loosely to cover all transformations of the form (2.1), with L proper ($\det L > 0$) and orthochronous ($L^0_{\ 0} > 0$).

Example 2.2 (Translation)

Here L is the identity. The two observers are at rest relative to each other, with their axes aligned, but in different locations and with different settings for their clocks.

Example 2.3 (Rotation)

If $T = 0$ and

$$L = \begin{pmatrix} 1 & 0 & 0 & 0 \\ 0 & \cos\theta & \sin\theta & 0 \\ 0 & -\sin\theta & \cos\theta & 0 \\ 0 & 0 & 0 & 1 \end{pmatrix}$$

then the two observers are at rest relative to each other at the same location, but their spatial axes are related by a rotation about the common z-axis.

Example 2.4 (Null rotation)

A less familiar Lorentz transformation is the *null rotation*

$$L = \frac{1}{2} \begin{pmatrix} 3 & 1 & 2 & 0 \\ -1 & 1 & -2 & 0 \\ 2 & 2 & 2 & 0 \\ 0 & 0 & 0 & 2 \end{pmatrix},$$

a combination of boost and rotation.

2.2 Inertial Coordinates

The extension of relativity to encompass gravitation requires the admission of more general transformations between space–time coordinate systems, in particular to allow for the relative acceleration of observers in free-fall.

Although we are still within the framework of special relativity, and the coordinates are still inertial, it will be helpful in making the transition to use notation in which the space and time coordinates are more explicitly on an equal footing. We therefore write

$$t = x^0, \qquad x = x^1, \qquad y = x^2, \quad \text{and} \quad z = x^3.$$

So the coordinates are labelled by upper indices. This is important, if unfamiliar: a lot of information will be stored by making a distinction between upper and lower indices.

With this notation, we can write (2.1) in the compact form

$$x^a = \sum_{b=0}^{3} L^a{}_b \tilde{x}^b + T^a \qquad (a = 0, 1, 2, 3). \tag{2.4}$$

Note that we keep track of the order of the indices on L. The upper index a comes first; it labels the rows of the matrix. The lower index b labels the columns, and comes second. By differentiating, we have that

$$L^a{}_b = \frac{\partial x^a}{\partial \tilde{x}^b}$$

and that

$$(L^{-1})^a{}_b = \frac{\partial \tilde{x}^a}{\partial x^b}.$$

Further notational economies are achieved by the adopting the following conventions and special notations.

The summation and range conventions

When an index is repeated in an expression (a *dummy* index), a sum over 0,1,2,3 is implied. An index that is not summed is a *free index*. Any equation is understood to hold for all possible values of its free indices. To apply the conventions consistently, an index must never appear more than twice in any term in an expression, once as an upper index and once as a lower index.

The metric coefficients and the Kronecker delta

We define the quantities g_{ab}, g^{ab}, and δ^a_b by

$$g_{ab} = g^{ab} = \begin{cases} 1 & a = b = 0 \\ -1 & a = b \neq 0 \\ 0 & \text{otherwise} \end{cases} \qquad \delta^a_b = \begin{cases} 1 & a = b \\ 0 & \text{otherwise} \end{cases}.$$

Later on, in general relativity, the 'metric coefficients' g_{ab} and g^{ab} will no longer be constant, nor will the coefficients with upper indices be the same as those with lower indices. On the other hand, the Kronecker delta δ^a_b will still be defined in this way.

The notation is very efficient; without it, calculations in relativity tend to be overwhelmed by a mass of summation signs. It does, however, have to be used with care and strict discipline. Free indices—indices for which there is no summation—must balance on the two sides of an equation. Excessive repetition can lead to ambiguous expressions in which it is not possible to restore the summation signs in a unique way. The following illustrate some of the uses and pitfalls of the notation.

Example 2.5

We can now omit the summation sign in (2.4). It becomes

$$x^a = L^a_{\ b}\tilde{x}^b + T^a. \tag{2.5}$$

Repetition of b implies summation over $0, 1, 2, 3$, and the range convention means that the equation is understood to hold as the free index a runs over the values $0, 1, 2, 3$.

Example 2.6

If two events have coordinates x^a and y^a in the first system and $\tilde{x}^a$ and $\tilde{y}^a$ in the second system, then

$$x^a - y^a = L^a_{\ b}(\tilde{x}^b - \tilde{y}^b) = L^a_{\ b}\tilde{x}^b - L^a_{\ b}\tilde{y}^b. \tag{2.6}$$

This illustrates that one must take care about what is meant by a 'term in an expression'. In principle, you should multiply out all the brackets before applying the summation rule; otherwise the threefold repetition of b in the middle expression could cause confusion. In practice, however, the meaning is clear, and the mild notational abuse in taking the summation through the brackets is accepted without causing difficulty.

Example 2.7

The Lorentz condition $L^{\mathrm{t}}gL = g$ becomes

$$L^c_{\ a}L^d_{\ b}g_{cd} = g_{cd}\frac{\partial x^c}{\partial \tilde{x}^a}\frac{\partial x^d}{\partial \tilde{x}^b} = g_{ab}.$$

Note that it does not matter in which order one writes the Ls and gs as long as the indices are 'wired up' correctly. In this equation a, b are free, whereas c, d are *dummy indices*, like dummy variables in an integral. The sum over c is the sum in the matrix product $L^{\mathrm{t}}g$, and the sum over d is the sum in the matrix product gL.

Similarly, $L^{-1}g^{-1}(L^{\mathrm{t}})^{-1} = g^{-1}$ becomes

$$g^{cd}\frac{\partial \tilde{x}^a}{\partial x^c}\frac{\partial \tilde{x}^b}{\partial x^d} = g^{ab}. \tag{2.7}$$

Example 2.8

If one combines two coordinate transformations

$$x^a = K^a{}_b \tilde{x}^b, \qquad \tilde{x}^a = L^a{}_b \hat{x}^b + T^a \tag{2.8}$$

then the result is

$$x^a = K^a{}_b L^b{}_c \hat{x}^c + K^a{}_b T^b. \tag{2.9}$$

To avoid ambiguity, it is necessary to change the dummy index in the second equation before making the substitution. It is then clear that there are two sums, over $b = 0, 1, 2, 3$ and over $c = 0, 1, 2, 3$. If you did not do this, then you would end up with the ambiguous expression $K^a{}_b L^b{}_b$, which could mean $\sum_{b=0}^{3} K^a{}_b L^b{}_b$.

Example 2.9

Written in full, the equation $A_a C^a = B_a C^a$ is

$$A_0 C^0 + A_1 C^1 + A_2 C^2 + A_3 C^3 = B_0 C^0 + B_1 C^1 + B_2 C^2 + B_3 C^3.$$

In the compact form, there is a temptation to cancel C^a to deduce that $A_a = C_a$. The full form shows that this temptation must be resisted.

Example 2.10

As a final illustration, we note that

$$g_{ab} g^{bc} = \delta^c_a. \tag{2.10}$$

Equivalently, $g_{..} g^{..}$ is the identity matrix; here $g_{..}$ and $g^{..}$ are the 4×4 matrices with, respectively, entries g_{ab} and g^{ab}. In (2.10), c, a are free indices and b is a dummy index. The same equation holds in general relativity, but there the metric coefficients are not constant.

2.3 Four-Vectors

A four-vector in special relativity has four components V^0, V^1, V^2, V^3. Under the change of coordinates (2.5), they transform by

$$\begin{pmatrix} V^0 \\ V^1 \\ V^2 \\ V^3 \end{pmatrix} = L \begin{pmatrix} \tilde{V}^0 \\ \tilde{V}^1 \\ \tilde{V}^2 \\ \tilde{V}^3 \end{pmatrix}, \tag{2.11}$$

That is, $V^a = L^a{}_b \tilde{V}^b$. The three Cartesian components of a vector $\boldsymbol{x}$ in Euclidean space behave in the same way. They change by

$$\begin{pmatrix} x_1 \\ x_2 \\ x_3 \end{pmatrix} = H \begin{pmatrix} \tilde{x}_1 \\ \tilde{x}_2 \\ \tilde{x}_3 \end{pmatrix}$$

when the axes are rotated by an orthogonal matrix H; and they are unchanged when the origin is translated.

Later on, we need to transform four-vector components under general coordinate transformations. So that we can carry over results from special relativity with the minimum of adaptation, we restate (2.11) by substituting $L^a{}_b = \partial x^a / \partial \tilde{x}^b$. Then the following definition is equivalent to the transformation rule in special relativity, and extends directly to the general theory.

Definition 2.11

A four-vector is an object with components V^a which transform by

$$V^a = \frac{\partial x^a}{\partial \tilde{x}^b} \tilde{V}^b$$

under change of inertial coordinates.

The only new feature when we come to allow general coordinate transformations will arise from the fact that the *Jacobian matrix* $\partial x^a / \partial \tilde{x}^b$ will not be constant, and so the transformation will vary from event to event: we shall have a distinct space of four-vectors at each event. Connecting them—that is, deciding when two vectors at different events are the same—is a central problem. We come to that later; for the moment all the coordinates are inertial and the Jacobian matrix is constant.

Example 2.12

The four-velocity: if $x^a = x^a(\tau)$ is the worldline of a particle, parametrized by proper time τ, then the four-velocity has components $V^a = \mathrm{d}x^a / \mathrm{d}\tau$. Under coordinate change

$$V^a = \frac{\mathrm{d}x^a}{\mathrm{d}\tau} = \frac{\partial x^a}{\partial \tilde{x}^b} \frac{\mathrm{d}\tilde{x}^b}{\mathrm{d}\tau}, \tag{2.12}$$

so the four-vector transformation rule is a consequence of the chain rule.

2.4 Tensors in Minkowski Space

Other objects in special relativity have similar transformation rules. *Tensor algebra* draws the various rules together into a common framework. The basic idea is that a set of physical quantities measured by one observer can be put together as the components of a single *tensor* in space–time. A four-vector is an example of a tensor. There is then a standard transformation rule that allows one to calculate the components in another coordinate system, and hence the same quantities as measured by a second observer. For example, the energy and momentum of a particle (in units with $c = 1$) form the time and space components of a four-vector. If they are known in one frame, then the transformation rule gives their values in another. Two other examples should be familiar.

Example 2.13

The components of the electric field $\boldsymbol{E}$ and the magnetic field $\boldsymbol{B}$ fit together to form the *electromagnetic* (EM) *field*

$$F = \begin{pmatrix} 0 & -E_1 & -E_2 & -E_3 \\ E_1 & 0 & -B_3 & B_2 \\ E_2 & B_3 & 0 & -B_1 \\ E_3 & -B_2 & B_1 & 0 \end{pmatrix} = \begin{pmatrix} F^{00} & F^{01} & F^{02} & F^{03} \\ F^{10} & F^{11} & F^{12} & F^{13} \\ F^{20} & F^{21} & F^{22} & F^{23} \\ F^{30} & F^{31} & F^{32} & F^{33} \end{pmatrix}, \quad (2.13)$$

which transforms by $F = L\tilde{F}L^t$. That is,

$$F^{ab} = L^a{}_c L^b{}_d \tilde{F}^{cd} = \frac{\partial x^a}{\partial \tilde{x}^c} \frac{\partial x^b}{\partial \tilde{x}^d} \tilde{F}^{cd}. \quad (2.14)$$

Example 2.14

The *gradient covector* of a function $f(x^a)$ of the space–time coordinates has components $\partial_a f$, where $\partial_a = \partial/\partial x^a$. These transform by the chain rule

$$\partial_a f = \frac{\partial \tilde{x}^b}{\partial x^a} \tilde{\partial}_b f. \quad (2.15)$$

Note that it is $\partial\tilde{x}/\partial x$ on the right-hand side, not $\partial x/\partial\tilde{x}$, so this is not the four-vector transformation rule, but rather a dual form of the rule. Hence the term 'covector'.

Definition 2.15

A tensor of type (p, q) is an object that assigns a set of components $T^{a\ldots b}{}_{c\ldots d}$ (p upper indices, q lower indices) to each inertial coordinate system, with the transformation rule under change of inertial coordinates

$$T^{a...b}{}_{c...d} = \frac{\partial x^a}{\partial \tilde{x}^e} \cdots \frac{\partial x^b}{\partial \tilde{x}^f} \frac{\partial \tilde{x}^h}{\partial x^c} \cdots \frac{\partial \tilde{x}^k}{\partial x^d} \tilde{T}^{e...f}{}_{h...k}.$$

A tensor can be defined at a single event, or along a curve, or on the whole of space–time, in which case the components are functions of the coordinates and we call T a tensor field. If $q = 0$ then there are only upper indices and the tensor is said to be *contravariant*; if $p = 0$, then there are only lower indices and the tensor is said to be *covariant*.

The definition is uncompromisingly pragmatic: a 'tensor' is defined in terms of the transformation rule for its components, leaving hanging the question of what, exactly, a tensor is. A four-vector can at least be pictured as an arrow in space–time, by analogy with a vector in space. A tensor with a large number of indices is not easily pictured as a geometric object, although this can be done with some ingenuity and willingness to lose contact with the physical context. More mathematically appealing definitions avoid this unease, but are not strictly necessary to get to grips with the theory; there is some discussion in the last chapter of [23]. One needs to become familiar with tensor algebra to do relativity, and this is best done by practice. Formal definitions and precise statements of the rules are not always helpful.

There is one serious point here that goes beyond the aesthetics of various characterizations of a 'tensor'. It should be checked that the transformation rule is consistent: that is, that in passing from coordinate system x^a to $\tilde{x}^a$ to $\hat{x}^a$, one gets the same transformation as by the direct route from x^a to $\hat{x}^a$. In fact, this follows from the product rule for Jacobian matrices

$$\frac{\partial x^a}{\partial \hat{x}^c} = \frac{\partial x^a}{\partial \tilde{x}^b} \frac{\partial \tilde{x}^b}{\partial \hat{x}^c}.$$

Example 2.16

A four-vector V^a is a tensor of type $(1, 0)$, also called a vector or contravariant vector.

Example 2.17

The gradient covector $\partial_a f$ is a tensor of type $(0, 1)$. A tensor α_a of type $(0, 1)$ is generally called a covector or covariant vector.

Example 2.18

The Kronecker delta is a tensor of type $(1, 1)$ because

$$\delta^c_d \frac{\partial x^a}{\partial \tilde{x}^c} \frac{\partial \tilde{x}^d}{\partial x^b} = \frac{\partial x^a}{\partial \tilde{x}^c} \frac{\partial \tilde{x}^c}{\partial x^b} = \delta^a_b \,, \tag{2.16}$$

by the chain rule.

Example 2.19

The *contravariant metric* has components g^{ab} and is a tensor of type $(2, 0)$, by (2.7). The *covariant metric* has components g_{ab} and is a tensor of type $(0, 2)$.

Both the Kronecker delta and the metric in Minkowski space are special in that they have the same components in every inertial frame. For a general tensor, the components in different frames are not the same.

As with four-vectors, the same definition will stand for general coordinate transformations, with the same caution that the transformation is then different at different events. The general strategy will be to identify tensors by their components in a 'local inertial frame' set up by an observer in free-fall, and then to use the transformation rule to find their components in other coordinate systems. The Kronecker delta will still have the same components in all systems, but the metric tensor will not.

2.5 Operations on Tensors

Addition

For S, T of the same type: $S + T$ has components

$$S^{a...b}{}_{c...d} + T^{a...b}{}_{c...d} \,.$$

Multiplication by scalars

A scalar at an event is simply a number. A scalar field is a function on space–time. The value of a scalar is unchanged by coordinate transformations. We can multiply a tensor T by a scalar f to get a tensor of the same type with components $f T^{a...b}{}_{c...d}$. The operations of addition and multiplication by constant scalars make the space of tensors of type (p, q) into a vector space of dimension 4^{p+q}.

Tensor product

It S, T are tensors of types (p, q), (r, s), respectively, then the *tensor product* is the tensor of type $(p + r, q + s)$ with components $S^{a...b}{}_{c...d} T^{e...f}{}_{g...h}$. It is denoted by ST or $S \otimes T$.

Differentiation

If T is a tensor field of type (p, q), then ∇T is defined to be the tensor of type $(p, q + 1)$ with components

$$\nabla_a T^{b...c}{}_{d...e} = \partial_a T^{b...c}{}_{d...e}, \qquad \partial_a = \frac{\partial}{\partial x^a}.$$

Under change of *inertial* coordinates,

$$\begin{aligned} \partial_a T^{b...}{}_{d...} &= \frac{\partial \tilde{x}^t}{\partial x^a} \tilde{\partial}_t \left[\frac{\partial x^b}{\partial \tilde{x}^r} \cdots \frac{\partial \tilde{x}^s}{\partial x^d} \cdots \tilde{T}^{r...}{}_{s...} \right] \\ &= \frac{\partial \tilde{x}^t}{\partial x^a} \frac{\partial x^b}{\partial \tilde{x}^r} \cdots \frac{\partial \tilde{x}^s}{\partial x^d} \cdots \tilde{\partial}_t \tilde{T}^{r...}{}_{s...}, \end{aligned}$$

which is the correct transformation rule for tensor components of type $(p, q + 1)$.

Note that we are still working in the context of special relativity: the calculation only works because $\partial x / \partial \tilde{x}$ is constant. We have to work harder to define differentiation in curved space–time.

Contraction

If T is of type $(p + 1, q + 1)$, then we can form a tensor S of type (p, q) by contracting on the first upper index and first lower index of T:

$$S^{b...c}{}_{e...f} = T^{ab...c}{}_{ae...f}.$$

Note that there is a sum over a. Under change of coordinates

$$\begin{aligned} S^{b...c}{}_{e...f} &= T^{ab...c}{}_{ae...f} \\ &= \frac{\partial x^a}{\partial \tilde{x}^k} \frac{\partial x^b}{\partial \tilde{x}^l} \cdots \frac{\partial x^c}{\partial \tilde{x}^m} \frac{\partial \tilde{x}^s}{\partial x^a} \frac{\partial \tilde{x}^t}{\partial x^e} \cdots \frac{\partial \tilde{x}^u}{\partial x^f} \tilde{T}^{kl...m}{}_{st...u} \\ &= \frac{\partial x^b}{\partial \tilde{x}^l} \cdots \frac{\partial x^c}{\partial \tilde{x}^m} \frac{\partial \tilde{x}^t}{\partial x^e} \cdots \frac{\partial \tilde{x}^u}{\partial x^f} \tilde{S}^{l...m}{}_{t...u} \end{aligned}$$

because

$$\frac{\partial x^a}{\partial \tilde{x}^k} \frac{\partial \tilde{x}^s}{\partial x^a} = \delta^s_k.$$

One can also contract on other pairs of indices, one upper and one lower.

Raising and lowering

If α is a covector and $U^a = g^{ab}\alpha_b$, then U is a four-vector, formed by tensor multiplication combined with contraction. We write α^a for U^a and call the operation 'raising the index'. Raising the index changes the signs of the 1,2,3 components, but leaves the first component unchanged. The reverse operation is 'lowering the index': $V_a = g_{ab}V^b$. One similarly lowers and raises indices on tensors by taking the tensor product with the covariant or contravariant metric and contracting, for example, $T^a_b = g_{bc}T^{ac}$. One must be careful to keep track of the order of the upper and lower indices because T^a_b and $T_b{}^a$ are generally distinct. Do not risk confusion by writing either as T^a_b.

Example 2.20

If f is scalar field, then $\nabla^a f$, where

$$(\nabla^a f) = (\partial_t f, -\partial_x f, -\partial_y f, -\partial_z f)$$

is a four-vector field. It is the 'gradient four-vector'.

Example 2.21

If U and V are four-vectors, then

$$g(U, V) = g_{ab}U^a V^b = U^a V_a = U_a V^a.$$

Example 2.22

Raising one index on g_{ab} or lowering one index on g^{ab} gives the Kronecker delta because $g^{ab}g_{bc} = \delta^a_c$.

Example 2.23

Suppose that $(S^a) = (1, 0, 0, 0)$ and $(T^a) = (1, 1, 0, 0)$. Then $S \otimes T$ and $T \otimes S$ have respective components

$$(S^a T^b) = \begin{pmatrix} 1 & 1 & 0 & 0 \\ 0 & 0 & 0 & 0 \\ 0 & 0 & 0 & 0 \\ 0 & 0 & 0 & 0 \end{pmatrix}, \qquad (T^a S^b) = \begin{pmatrix} 1 & 0 & 0 & 0 \\ 1 & 0 & 0 & 0 \\ 0 & 0 & 0 & 0 \\ 0 & 0 & 0 & 0 \end{pmatrix}.$$

Note that $S \otimes T \neq T \otimes S$, but when written as matrices, as above, the components of $S \otimes T$ and $T \otimes S$ are related by transposition.

EXERCISES

2.1. For each of the following, either write out the equation with the summation signs included explicitly or say in a few words why the equation is ambiguous or does not make sense.

   (i) $x^a = L^a{}_b M^b{}_c \hat{x}^c$.

   (ii) $x^a = L^b{}_c M^c{}_d \hat{x}^d$.

   (iii) $\delta^a_b = \delta^a_c \delta^c_d \delta^d_b$.

   (iv) $\delta^a_b = \delta^a_c \delta^c_c \delta^c_b$.

   (v) $x^a = L^a{}_b \hat{x}^b + M^a{}_b \hat{x}^b$.

   (vi) $x^a = L^a{}_b \hat{x}^b + M^a{}_c \hat{x}^c$.

   (vii) $x^a = L^a{}_c \hat{x}^c + M^b{}_c \hat{x}^c$.

2.2. Show that for any tensors S, T, U, with T and U of the same type, $S \otimes (T + U) = S \otimes T + S \otimes U$.

2.3. The *alternating symbol* is defined by

$$\varepsilon_{abcd} = \begin{cases} 1 & \text{if } abcd \text{ is an even permutation of 0123} \\ -1 & \text{if } abcd \text{ is an odd permutation of 0123} \\ 0 & \text{otherwise.} \end{cases}$$

Show that if T, X, Y, Z are four-vectors with $T = (1, 0)$, $X = (0, \boldsymbol{x})$, $Y = (0, \boldsymbol{y})$, and $Z = (0, \boldsymbol{z})$, then

$$\varepsilon_{abcd} T^a X^b Y^c Z^d = \boldsymbol{x}.(\boldsymbol{y} \wedge \boldsymbol{z}) \,.$$

2.4. Let ε have components ε_{abcd} in every inertial coordinate system.

   (i) Show that ε is a tensor of type $(0, 4)$.

   (ii) Write down the values of the components of the contravariant tensor ε^{abcd}.

   (iii) Show that $\varepsilon_{abcd} \varepsilon^{abcd} = -24$ and that $\varepsilon_{abcd} \varepsilon^{abce} = -6\delta^e_d$.

2.5. Maxwell's equations are

$$\begin{aligned} \text{div}\,\boldsymbol{E} &= \epsilon_0^{-1} \rho \\ \text{div}\,\boldsymbol{B} &= 0 \\ \text{curl}\,\boldsymbol{B} - \partial_t \boldsymbol{E} &= \mu_0 \boldsymbol{J} \\ \text{curl}\,\boldsymbol{E} + \partial_t \boldsymbol{B} &= 0 \,, \end{aligned}$$

where $\epsilon_0\mu_0 - 1$ in these units in which $c - 1$. Show that they take the tensor form

$$\partial_a F^{ab} = \epsilon_0^{-1} J^b \qquad \text{and} \qquad \partial_a F_{bc} + \partial_b F_{ca} + \partial_c F_{ab} = 0\,,$$

where $J = (\rho, \boldsymbol{J})$ is the current four-vector.

2.6. Let F^{ab} be an electromagnetic field tensor. Write down the components of the dual tensor $F^*_{ab} = \frac{1}{2}\varepsilon_{abcd}F^{cd}$ in terms of the components of the electric and magnetic fields. By considering the scalars $F_{ab}F^{ab}$ and $F_{ab}F^{*ab}$, show that $\boldsymbol{E}\,.\,\boldsymbol{B}$ and $\boldsymbol{E}\,.\,\boldsymbol{E} - \boldsymbol{B}\,.\,\boldsymbol{B}$ are invariants.

2.7. An observer moves through an electromagnetic field F^{ab} with four-velocity U^a. Show that $U^aU_a = 1$. Show that the observer sees no magnetic field if $F^{*ab}U_b = 0$, and show that this equation is equivalent to

$$\boldsymbol{B}\,.\,\boldsymbol{u} = 0 \qquad \text{and} \qquad \boldsymbol{B} - \boldsymbol{u} \wedge \boldsymbol{E} = 0\,.$$

Hence show that there exists a frame in which the magnetic field vanishes at an event if and only if in every frame $\boldsymbol{E}\,.\,\boldsymbol{B} = 0$ and $\boldsymbol{B}\,.\,\boldsymbol{B} < \boldsymbol{E}\,.\,\boldsymbol{E}$ at the event.

3

Energy-Momentum Tensors

Einstein's general theory has at its heart an equation that, like Poisson's equation, relates the gravitational field of a distribution of matter to its energy density. The quantity that encodes energy density in special relativity is a symmetric two-index tensor called the *energy-momentum tensor*. We introduce it first in the simplest case of a noninteracting distribution of particles, and then extend the definition to fluids and to electromagnetic fields.

3.1 Dust

Consider a cloud of particles ('dust'), in which the velocities of the individual particles vary smoothly from event to event and from time to time. There is one worldline through each event and the four-velocities of the individual dust particles make up a four-vector field U. For the moment, we suppose that there are no external forces or interactions, so each particle moves in a straight line at constant speed.

We now address the question: what is the energy density seen by an observer moving through the dust with four-velocity V? The observer's worldline is the dashed line in Figure 3.1. The answer depends on V because

(i) The energy of each individual particle depends on its velocity relative to the observer; and

(ii) Moving volumes appear to contract.

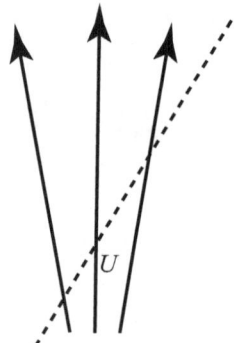

Figure 3.1 A 'dust' cloud

The answer is important in general relativity because it involves the introduction of the *energy-momentum tensor*, which is the 'source term' in Einstein's equations, analogous to the current four-vector in Maxwell's equations.

Definition 3.1

The rest density ρ is a scalar. It is defined at an event A to be the rest mass per unit volume measured in a frame in which the particles at A are at rest. If there are n particles per unit volume in this frame and each has rest mass m, then $\rho = nm$.

Consider the particles that occupy a unit volume at an event A in the rest frame of the particles at A. Suppose that in this frame the observer is moving along the negative x-axis with speed v. To the observer, each particle at A appears to have velocity $(v, 0, 0)$ and to have energy

$$m\gamma(v) = \frac{m}{\sqrt{1 - v^2}}\,.$$

The particles appear to occupy a volume $1/\gamma(v) = \sqrt{1 - v^2}$. Therefore the observer measures the energy density to be $\gamma(v)^2\rho$.

Definition 3.2

The energy-momentum tensor of the dust cloud is the tensor field with components $T^{ab} = \rho U^a U^b$. It is a tensor of type $(2, 0)$ because it is the tensor product of two four-vectors, multiplied by a scalar.

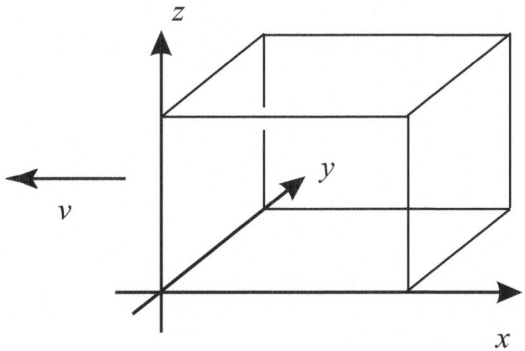

Figure 3.2 The transformation of density

Proposition 3.3

The energy density measured by an observer moving through the cloud with four-velocity V is $\rho_V = T_{ab}V^aV^b$.

Proof

In the rest frame of the observer,

$$(V^a) = (1,0,0,0) \qquad (U^a) = \gamma(v)(1,v,0,0).$$

Therefore $T_{ab}V^aV^b = \rho(U_aV^a)^2 = \rho\gamma(v)^2$. $\square$

Thus the 00-component of the energy-momentum tensor in the observer's rest frame is the energy density. What about the other components? Consider the four-vector $T^{ab}V_b$. Its temporal component in the observer's frame is ρ_V. Its spatial part is $\boldsymbol{f} = \rho_V\boldsymbol{u}$, where $\boldsymbol{u}$ is the particle velocity relative to the observer. This represents the *energy flow*. The particles that cross a small surface element $\mathrm{d}S$ with normal $\boldsymbol{n}$ in the observer's time δt occupy a volume $\boldsymbol{u}.\boldsymbol{n}\,\mathrm{d}S\,\delta t$ after they cross. The total energy of these particles as measured by the observer is therefore $\boldsymbol{f}.\boldsymbol{n}\,\mathrm{d}S$. So if Ω is a fixed volume in the observer's frame, with boundary surface $\partial\Omega$ and outward pointing normal, then conservation of energy requires that

$$\frac{\mathrm{d}}{\mathrm{d}t}\int_{\Omega}\rho_V\,\mathrm{d}V + \int_{\partial\Omega}\boldsymbol{f}.\mathrm{d}\boldsymbol{S} = 0\,. \tag{3.1}$$

The surface integral represents the total rate at which energy is flowing out of Ω. By taking the time derivative under the first integral sign and by applying the divergence theorem to the surface integral, we get

$$\int_\Omega \left(\frac{\partial \rho_V}{\partial t} + \operatorname{div} \boldsymbol{f} \right) \mathrm{d}V = 0 \,.$$

Because this holds for any fixed volume, we have the *continuity equation*

$$\frac{\partial \rho_V}{\partial t} + \operatorname{div} \boldsymbol{f} = 0 \,. \tag{3.2}$$

Equivalently,

$$\nabla_a (T^{ab} V_b) = 0 \,.$$

Because V is constant and the second equation holds for any observer, it follows that

$$\nabla_a T^{ab} = 0 \,. \tag{3.3}$$

Thus we see that conservation of energy for all inertial observers is equivalent to (3.3). If the dust particles are moving slowly relative to the observer, then $\rho_V \sim \rho$ and (3.2) reduces to the classical continuity equation of fluid dynamics.

We can also deduce the equation of motion of the individual dust particles from the conservation law (3.3). If we substitute $T^{ab} = \rho U^a U^b$, then we obtain

$$\rho U^a \nabla_a U^b = -U^b \nabla_a (\rho U^a) \,.$$

So $U^a \nabla_a U^b$ is parallel to U^b. On the other hand, $U_b U^b = 1$, and so

$$0 = U^a \nabla_a (U^b U_b) = 2 U_b U^a \nabla_a U^b \,,$$

which implies that $U^a \nabla_a U^b$ is also orthogonal to U^b. Consequently

$$\frac{\mathrm{d}U^b}{\mathrm{d}\tau} = U^a \nabla_a U^b = 0 \,,$$

where τ is the proper time along a particle worldline. In other words, U^a is constant along each particle worldline, and so the individual particles move in straight lines at constant speeds. This takes us back to where we started, but the point is that the equation of motion is determined by the requirement that energy measured by any inertial observer should be conserved.

3.2 Fluids

The definition extends to a general relativistic fluid. We picture a fluid as a large number of superimposed streams of particles with different velocities. Each stream has its own energy-momentum tensor, and their sum T^{ab} encodes the energy density for the whole fluid. An inertial observer with four-velocity V measures energy density $\rho_V = T^{ab}V_aV_b$, and sees an energy flow given by the spatial part of the four-vector $T^{ab}V_b$. The different streams interact through collisions, but energy is conserved in the rest frame of an inertial observer, so the same energy conservation argument as before, applied to a fixed volume in an observer's frame, gives $\nabla_a(T^{ab}V_b) = 0$. This holds for the four-velocity V^a of any observer, so as before we have

$$\nabla_a T^{ab} = 0\,.$$

How does such a fluid acquire a well-defined bulk velocity? It is through the existence of a frame at each event in which the energy density is minimal.

The energy density measured at some event by an observer moving with velocity v through a stream of particles with velocity u and rest density ρ is

$$\rho U^a U^b V_a V_b = \rho\gamma(u)\gamma(v)(1 - \boldsymbol{u}.\boldsymbol{v})^2\,.$$

As $v \to 1$, therefore, the observed density tends to infinity. Because each individual stream has positive density, the same must be true of the whole fluid. So if we put $V = \gamma(v)(1, \boldsymbol{v})$, and regard

$$\rho_V = T^{ab}V_aV_b$$

as a function of $\boldsymbol{v}$, then ρ_V is positive whenever $|\boldsymbol{v}| < 1$ and $\rho_V \to \infty$ as $|\boldsymbol{v}| \to 1$. Consequently ρ_V must achieve its minimum for some value of $\boldsymbol{w}$ of $\boldsymbol{v}$.

Consider the corresponding four-velocity W^a. By the following argument, we can characterize W^a as the unique timelike eigenvector of T^{ab}. Let X^a be a four-vector orthogonal to W^a; that is, $W^aX_a = 0$. Suppose that the components of X^a are small. If we ignore quadratic terms in these small quantities, then $W^a + X^a$ is also a four-velocity because

$$(W^a + X^a)(W_a + X_a) = W^aW_a + 2W^aX_a = 1\,.$$

With the same approximation, we also have

$$\begin{aligned}
T^{ab}(W_a + X_a)(W_b + X_b) &= T^{ab}W_aW_b + 2T^{ab}W_aX_b \\
&\geq T^{ab}W_aW_b\,.
\end{aligned}$$

Therefore

$$T^{ab}W_aX_b \geq 0$$

for small X^a. But this must still hold if we replace X^a by $-X^a$, so we deduce that

$$T^{ab}W_aX_b = 0 \, .$$

Because this is true for any X^a orthogonal to W^a, it follows that $T^{ab}W_a$ is parallel to W^b, and thus that W^b is an eigenvector of the energy-momentum tensor. That is,

$$T^{ab}W_a = \rho W^b$$

for some scalar ρ. By contracting with W_b, we see that ρ is the minimum possible value of ρ_V.

Definition 3.4

The four-velocity W^a that satisfies the eigenvector equation $T^{ab}W_a = \rho W^b$ at some event is the *rest-velocity* of the fluid at the event, and the corresponding eigenvalue ρ is the *rest density*.

Exercise 3.1

Show that the rest-velocity at an event is unique.

A rest frame of the fluid at an event is a frame in which $(W^a) = (1, 0, 0, 0)$ and in which the components of the matrix (T^{ab}) can be written in block form

$$(T^{ab}) = \begin{pmatrix} \rho & 0 \\ 0 & \sigma \end{pmatrix} \, ,$$

where σ is a 3×3 matrix. In general, the σ has three distinct eigenvectors and these pick out three special directions in the fluid. A *perfect fluid* is one for which there are no special directions and therefore one for which σ is a multiple of the identity. Such a fluid is *isotropic*: it looks the same in every direction at the event. For an isotropic fluid, we have

$$T^{ab} = \rho W^a W^b - p(g^{ab} - W^a W^b)$$

for some scalar field p. If we expand the conservation law $\nabla_a T^{ab} = 0$ in a general inertial coordinate system, then this time we obtain

$$W^a \nabla_a \rho + (\rho + p) \nabla_a W^a = 0$$

and

$$(\rho + p) W^a \nabla_a W^b = (g^{ab} - W^a W^b) \nabla_a p \, .$$

If all the individual particle streams are moving with velocity much less than that of light, then the fluid velocity $\boldsymbol{w}$ will be small and p will be very much less than p. We can approximate four-velocity of the fluid by $(1, \boldsymbol{w})$, and ignore quadratic terms w^2 and pw. Our conservation equations then reduce to

$$\partial_t \rho + \boldsymbol{\nabla} \cdot (\rho \boldsymbol{w}) = 0, \qquad \rho\, \partial_t \boldsymbol{w} + \rho(\boldsymbol{w} \cdot \boldsymbol{\nabla})\boldsymbol{w} = -\boldsymbol{\nabla}p\,,$$

which are the continuity equation and Euler equation of nonrelativistic fluid dynamics. Thus in general we should interpret p as the pressure of the perfect fluid.

3.3 Electromagnetic Energy-Momentum Tensor

A second extension takes account of electromagnetic forces and of the energy carried by an electromagnetic field. Let us return to the case of a single stream of particles, but suppose now that the particles are charged, and that they interact electromagnetically, but are not subject to other external forces. If each particle has rest mass m and charge e, then the current four-vector at an event is $J = neU$, where n is the number of particles per unit volume in the rest frame of the particles at the event and V is their four-velocity. It has spatial part $\boldsymbol{J} = ne\gamma(u)\boldsymbol{u}$.

In the coordinates of an inertial observer with four-velocity V^a, the motion of each particle is governed by the Lorentz force law

$$m\frac{\mathrm{d}V^a}{\mathrm{d}\tau} = eF^{ab}V_b\,.$$

Hence it satisfies

$$\frac{\mathrm{d}}{\mathrm{d}t}\Big(m\gamma(u)\Big) = e\boldsymbol{E} \cdot \boldsymbol{u}\,.$$

It follows that between t and $t + \delta t$, the energy $m\gamma(u)$ of the particle changes by $e\boldsymbol{E} \cdot \boldsymbol{u}\, \delta t$. There are $n\gamma(u)$ particles per unit volume in the observer's frame. So the conservation equation for a volume Ω is now

$$\frac{\mathrm{d}}{\mathrm{d}t}\int_\Omega \rho_V \,\mathrm{d}V + \int_{\partial\Omega} \boldsymbol{f}.\mathrm{d}S = \int_\Omega ne\gamma(u)\boldsymbol{E} \cdot \boldsymbol{u}\,\mathrm{d}V\,.$$

But the right-hand side is

$$\int_\Omega \boldsymbol{E} \cdot \boldsymbol{J}\,\mathrm{d}V = \int_\Omega \frac{1}{\mu_0}\boldsymbol{E} \cdot \left(\operatorname{curl}\boldsymbol{B} - \frac{\partial\boldsymbol{E}}{\partial t}\right)\mathrm{d}V$$

by Maxwell's equations; see Exercise 2.5. Moreover

$$\boldsymbol{E} \cdot \operatorname{curl}\boldsymbol{B} = \operatorname{div}(\boldsymbol{B} \wedge \boldsymbol{E}) + \boldsymbol{B} \cdot \operatorname{curl}\boldsymbol{E} = \operatorname{div}(\boldsymbol{B} \wedge \boldsymbol{E}) - \boldsymbol{B} \cdot \frac{\partial\boldsymbol{B}}{\partial t}\,.$$

Hence

$$\frac{\mathrm{d}}{\mathrm{d}t}\int_\Omega\left(\rho_V + \tfrac{1}{2}\epsilon_0(\boldsymbol{E}\,.\,\boldsymbol{E} + \boldsymbol{B}\,.\,\boldsymbol{B})\right)\mathrm{d}V + \int_{\partial\Omega}(\boldsymbol{f} + \epsilon_0\boldsymbol{E}\wedge\boldsymbol{B})\,.\,\mathrm{d}\boldsymbol{S} = 0\,,$$

where ρ_V and $\boldsymbol{f}$ are as in (3.1). It makes sense, therefore, to identify the quantity[1]

$$\frac{\epsilon_0}{2}\boldsymbol{E}\,.\,\boldsymbol{E} + \frac{1}{2\mu_0}\boldsymbol{B}\,.\,\boldsymbol{B}$$

with the *energy density* of the electromagnetic field and to identify the vector

$$\frac{\boldsymbol{E}\wedge\boldsymbol{B}}{\mu_0}$$

with the *energy flux*. This vector is called the *Poynting vector*. The energy density and the Poynting vector are the temporal and spatial components of $\tau^{ab}V_b$, where

$$\tau^{ab} = \epsilon_0(F^{ac}F_c{}^b + \tfrac{1}{4}g^{ab}F_{cd}F^{cd})$$

is the *electromagnetic energy-momentum tensor*. Our conservation equation is now

$$\nabla_a(\rho U^a U^b + \tau^{ab}) = 0\,.$$

Neither the energy-momentum tensor of the particles nor that of the electromagnetic field is conserved on its own; but the combination is, as common sense and physical law demand.

EXERCISES

3.2. Show that the electromagnetic energy momentum tensor is symmetric.

3.3. Let τ^{ab} be the energy-momentum tensor of an electromagnetic field F. Show that

$$\tau^{ab} = \tfrac{1}{2}\epsilon_0\left(F^a{}_c F^{cb} + F^{*a}{}_c F^{*cb}\right)\,.$$

3.4. Show that, except when $F_{ab}F^{ab} = F_{ab}F^{*ab} = 0$, there are two independent real null four-vectors L such that $K^{ab}L_b = \lambda L^a$ for some λ. They are called the *principal null vectors*. Explain why this implies that the electromagnetic field does not have a unique 'bulk velocity'. How many principal null vectors are there when $F_{ab}F^{ab} = F_{ab}F^{*ab} = 0$? How are they related to the Poynting vector?

---

[1] In units in which $c = 1$, we have $\mu_0 = \epsilon_0^{-1}$, so one constant is redundant. We use both here simply to bring the definitions closer to their conventional form.

3.5. Show that for a perfect fluid, the conservation equation $\nabla_a T^{ab} = 0$ is equivalent to

$$\nabla_a(\rho W^a) + p\nabla_a W^a = 0, \qquad (\rho+p)\frac{\mathrm{d}W^a}{\mathrm{d}\tau} + (W^a W^b - g^{ab})\nabla_b p = 0\,,$$

where τ is the proper time along the worldlines of the fluid elements. Why does $\nabla_a(\rho W^a)$ not vanish?

4
Curved Space–Time

We are now ready to make the transition from Minkowski's space–time of special relativity to the curved space–time of general relativity. We build on two foundations: first, the equivalence principle, the local equivalence of the effects of acceleration and gravity, and second, the well-established apparatus of special relativity theory, applied over short times and small distances in free-fall. Our starting point is the following.

(GR1) Special relativity holds over small distances and short times in frames in free-fall, that is, in local inertial frames. In such frames we can set up local inertial coordinates as in Minkowski space.

(GR2) Gravity appears as the relative acceleration of nearby local inertial frames.

4.1 Local Inertial Frames

In special relativity, an inertial observer sets up an inertial coordinate system t, x, y, z by using Milne's radar method and by measuring the direction of propagation of light arriving from events at other locations. Two such systems are related by an inhomogeneous Lorentz transformation. If A and A' are two events with respective coordinates t, x, y, z and t', x', y', z' then the quantity

$$\sigma(A, A') = (t' - t)^2 - (x' - x)^2 - (y' - y)^2 - (z' - z)^2 \qquad (4.1)$$

is independent of the choice of coordinate system. It is called the *world function*; it depends only on the two events A, A'.

If $\sigma(A, A')$ is positive, then it is the square of the time interval from A to A' measured in a frame in which A and A' happen in the same place. If it is negative, then it is minus the square of the distance between A and A', measured in a frame in which they happen at the same time. If it is zero, then A and A' lie on the worldline of a photon.

In the presence of gravity, an observer in free-fall with worldline ω can set up *local inertial coordinates* in the same way, taking an event on ω as origin. The times and distances of other events are measured by the radar method, and the events' coordinates are found by adding information about direction of travel of the returning light signals. By GR1, all observers in free-fall will measure the same value of the world function for two nearby events. So if A is the origin and B is a nearby event with coordinates $\mathrm{d}t$, $\mathrm{d}x$, $\mathrm{d}y$, and $\mathrm{d}z$, then

$$\mathrm{d}s^2 = \mathrm{d}t^2 - \mathrm{d}x^2 - \mathrm{d}y^2 - \mathrm{d}z^2$$

is the same in all local inertial coordinate systems with origin A provided that we ignore third-order terms in the small quantities $\mathrm{d}t, \mathrm{d}x, \mathrm{d}y, \mathrm{d}z$. Although it is conventional to write it as a square, $\mathrm{d}s^2$ can be positive, negative, or zero. It has the same interpretation as in special relativity.

Timelike separation. If $\mathrm{d}s^2 > 0$, then $\mathrm{d}s$ is the time from A to B on a clock travelling between the two events in free-fall.

Null separation. If $\mathrm{d}s^2 = 0$, then A and B lie on the worldline of a photon.

Spacelike separation. If $\mathrm{d}s^2 < 0$, then $\mathrm{d}s^2 = -D^2$, where D is the distance from A to B measured in a frame in free-fall in which A and B are simultaneous.

The change from special relativity is that the interpretation of $\mathrm{d}s^2$ is now an approximation, valid when A is the origin of the coordinate system set up by the free-falling observer and B is nearby, and valid only to the extent that the coordinates of B can be treated as small quantities.

A second application of GR1 gives the equations of motion of particles in free-fall, either massive particles moving at less than the velocity of light or photons moving at the velocity of light. Their worldlines are defined by expressing t, x, y, z as functions of a parameter τ. In special relativity, τ is proper time in the case of a particle with mass—that is, the time measured by a clock moving with the particle—or an affine parameter in the case of a photon. Either way,

$$\frac{\mathrm{d}^2 t}{\mathrm{d}\tau^2} = \frac{\mathrm{d}^2 x}{\mathrm{d}\tau^2} = \frac{\mathrm{d}^2 y}{\mathrm{d}\tau^2} = \frac{\mathrm{d}^2 z}{\mathrm{d}\tau^2} = 0 \,. \tag{4.2}$$

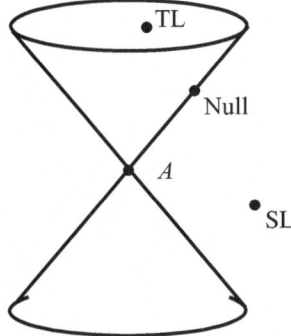

Figure 4.1 The displacement from A to B in the three cases

That is, the worldline is a straight line in space–time and the parameter is linear. In the presence of gravity, these equations must still hold at the origin of a local inertial coordinate system, but we do not expect them to hold at other events because the particle will acquire a small acceleration relative to the observer as it travels away from the origin. Thus we have the following.

Motion in free-fall. In free-fall, the motion of a particle satisfies (4.2) at any event A on the worldline in any local inertial coordinate system with origin A. In the case of a massive particle, τ is the time measured by a clock falling with the particle. In the case of a photon τ is an affine parameter.

We show that this is enough to determine the motion in general coordinates. By 'free-fall' is meant 'subject to no forces other than gravity'.

The coordinates t, x, y, z can only be used in the immediate neighbourhood of the origin. If we want to see what is happening at other events, then we must use a different coordinate system. So we now translate our conclusions thus far into general coordinates. As always, we want to keep in mind the analogy with mapmaking. The local inertial coordinates are analogous to the x, y coordinates on a large-scale map of a small area of the earth's surface. In that context, the distance between two nearby points is

$$\mathrm{d}s^2 = \mathrm{d}x^2 + \mathrm{d}y^2 \,,$$

where $\mathrm{d}x$ and $\mathrm{d}y$ are the differences in their x coordinates and in their y coordinates. Straight lines on the surface correspond to straight lines on the map, and there is a constant scale. But we need a different map for a different region: because of the curvature of the earth, we cannot construct a map of a large region with these properties. On a global scale, we must use a projection that distorts the local geometry in some way, and we can no longer compute the distance between two widely separated points by measuring their x and y coordinates on the map, and by applying Pythagoras's theorem.

Local inertial coordinate systems are analogous to large-scale maps. They can only be used to explore the immediate neighbourhood of an event. One can study a larger region of space–time by using a general coordinate system, but at the price of having a more complicated formula for the time and distance separation between nearby events. The geometry no longer looks like the flat geometry of Minkowski space.

A general coordinate system x^a on space–time is simply a labelling of events by four parameters. We should not think of the coordinates as having a direct interpretation in terms of the measurement of physical quantities. They are simply labels. Near the origin A of a local inertial coordinate system, t, x, y, z are functions of the x^as, so

$$\mathrm{d}t = \frac{\partial t}{\partial x^a}\, \mathrm{d}x^a + \text{second-order terms in } \mathrm{d}x$$

and so on. If we ignore third-order terms in the $\mathrm{d}x^a$s, then

$$\mathrm{d}s^2 = g_{ab}\mathrm{d}x^a\mathrm{d}x^b \tag{4.3}$$

at A, where

$$g_{ab} = \frac{\partial t}{\partial x^a}\frac{\partial t}{\partial x^b} - \frac{\partial x}{\partial x^a}\frac{\partial x}{\partial x^b} - \frac{\partial y}{\partial x^a}\frac{\partial y}{\partial x^b} - \frac{\partial z}{\partial x^a}\frac{\partial z}{\partial x^b}. \tag{4.4}$$

In an extension of our previous terminology, the coefficients $g_{ab} = g_{ba}$ are called the *metric coefficients*. Because $\mathrm{d}s^2$ is given by the same expression in all local inertial coordinate systems, the value of the right-hand side of (4.4) at A is independent of the choice of the local inertial coordinates t, x, y, z at A.

We can do a similar transformation to local inertial coordinates near any other event. So (4.3) holds throughout the region covered by the coordinates x^a. However, in general the metric coefficients g_{ab} vary from event to event. In contrast to the special theory, they are now dependent on the choice of space–time coordinates x^a.

If we replace the x^as by new coordinates $\tilde{x}^a$, then

$$\mathrm{d}s^2 = g_{ab}\mathrm{d}x^a\mathrm{d}x^b = \left(g_{cd}\frac{\partial x^c}{\partial \tilde{x}^a}\frac{\partial x^d}{\partial \tilde{x}^b} \right)\mathrm{d}\tilde{x}^a\mathrm{d}\tilde{x}^b.$$

So in the new coordinate system the metric coefficients are

$$\tilde{g}_{ab} = g_{cd}\frac{\partial x^c}{\partial \tilde{x}^a}\frac{\partial x^d}{\partial \tilde{x}^b}$$

or in matrix notation

$$\tilde{g} = J^t g J \qquad \text{where} \qquad J = \left(\frac{\partial x^a}{\partial \tilde{x}^b} \right).$$

A general real symmetric matrix can always be reduced to a diagonal matrix with diagonal entries ± 1 by a transformation $g \mapsto J^t g J$ for some matrix J. The diagonal form is determined by the signature, that is, by the signs of the eigenvalues. In the case of the matrix $g = (g_{ab})$ of metric coefficients, we know that we can reduce g to the diagonal matrix with diagonal entries $1, -1, -1, -1$ at any one event by transforming to local inertial coordinates at that event. Therefore the matrix g has one positive and three negative eigenvalues, which is usually expressed by saying that the metric has signature $+ - --$.

To summarize, in an arbitrary coordinate system, if $\mathrm{d}x^a$ is the coordinate separation between two nearby events A and B, then, to the second order in $\mathrm{d}x^a$,

$$\mathrm{d}s^2 = g_{ab} \mathrm{d}x^a \mathrm{d}x^b,$$

where the metric coefficients are evaluated at A and $\mathrm{d}s$ has the interpretation above. The coefficients g_{ab} have the following properties.

(MC1) They are smooth functions of the coordinates x^a.

(MC2) They are symmetric $g_{ab} = g_{ba}$.

(MC3) The matrix (g_{ab}) has signature $+ - --$ at every event.

(MC4) The metric coefficients transform under general coordinate transformations by

$$\tilde{g}_{ab} = g_{cd} \frac{\partial x^c}{\partial \tilde{x}^a} \frac{\partial x^d}{\partial \tilde{x}^b}.$$

Example 4.1

Suppose that $x^0 = t$, $x^1 = r$, $x^2 = \theta$, $x^3 = \varphi$, and

$$\mathrm{d}s^2 = \mathrm{d}t^2 - \mathrm{d}r^2 - r^2 \mathrm{d}\theta^2 - r^2 \sin^2 \theta \mathrm{d}\varphi^2. \tag{4.5}$$

Then we can reduce $\mathrm{d}s^2$ to the form $\mathrm{d}t^2 - \mathrm{d}x^2 - \mathrm{d}y^2 - \mathrm{d}z^2$ by the coordinate change $t = t$, $x = r \sin \theta \cos \varphi$, $y = r \sin \theta \sin \varphi$, $z = r \cos \theta$. So this is just the metric of special relativity in a noninertial coordinate system (spherical polars). We cannot reduce a general metric to the Minkowski form by a coordinate transformation. However, we can do it up to the second order in the coordinates at any one event, as we show in the next section.

It is conventional to specify the metric coefficients in general coordinates by giving an (infinitesimal) expression, referred to as the *metric*, for $\mathrm{d}s^2$ in the form

$$\mathrm{d}s^2 = g_{ab}\mathrm{d}x^a\mathrm{d}x^b\,.$$

For example, for Minkowski space in spherical polar coordinates, we read off from (4.5) that $g_{00} = 1$ and $g_{33} = -r^2\sin^2\theta$.

4.2 Existence of Local Inertial Coordinates

The central idea of general relativity is that a gravitational field can be described by a metric

$$\mathrm{d}s^2 = g_{ab}\,\mathrm{d}x^a\,\mathrm{d}x^b\,,$$

where the metric coefficients satisfy (MC1)–(MC4). In order to understand how such a metric can carry nontrivial information about gravity and how its coefficients can be interpreted in terms of observations made in free-fall, we explore the recovery from g_{ab} of local inertial coordinates. We show that these can always be found at any event in space-time, but that a general metric cannot be reduced globally to the Minkowski form by a change of coordinates. A general metric is not simply the metric of Minkowski space disguised by a coordinate transformation, as in the last example.

The recovery of local inertial coordinates begins with the following proposition.

Proposition 4.2

Let g_{ab} be a set of metric coefficients such that (MC1)–(MC4) hold and let A be the event $x^a = 0$. Then there exists a coordinate system $\tilde{x}^a$ such that $\tilde{x}^a = 0$ and $\tilde{\partial}_c\tilde{g}_{ab} = 0$ at A.

Proof

Define new coordinates $\tilde{x}^a$ by $x^a = \tilde{x}^a - \frac{1}{2}\Gamma^a_{bc}\tilde{x}^b\tilde{x}^c$, where the Γ^a_{bc}s are constants such that $\Gamma^a_{bc} = \Gamma^a_{cb}$. Let h_{ab} and k_{cab} denote, respectively, the values of g_{ab} and $\partial_c g_{ab}$ at the origin $x^a = 0$. Then, by Taylor's theorem,

$$g_{ab} = h_{ab} + x^c k_{cab} + O(2),$$

where '$O(2)$' denotes quadratic and higher-order terms in the x^as. It follows that

$$\tilde{g}_{ab} \;=\; g_{cd}\frac{\partial x^c}{\partial\tilde{x}^a}\frac{\partial x^d}{\partial\tilde{x}^b}$$

$$= (h_{cd} + x^m k_{mcd})(\delta_a^c - \Gamma^c_{ae}\tilde{x}^e)(\delta_b^d - \Gamma^d_{bf}\tilde{x}^f) + O(2)$$
$$= h_{ab} + \tilde{x}^c(k_{cab} - \Gamma_{abc} - \Gamma_{bac}) + O(2),$$

where $\Gamma_{abc} = h_{ad}\Gamma^d_{bc}$. We have used $x^a = \tilde{x}^a + O(2)$, as well as changing the labelling of the dummy indices. We want to choose $\Gamma_{abc} = \Gamma_{acb}$ so that

$$k_{cab} = \Gamma_{abc} + \Gamma_{bac}.$$

By permuting the indices, we would then also have

$$k_{bca} = \Gamma_{cab} + \Gamma_{acb}$$
$$k_{abc} = \Gamma_{bca} + \Gamma_{cba}.$$

By adding the first two of these and subtracting the third, we would then have that

$$\Gamma_{abc} = \tfrac{1}{2}(k_{cab} + k_{bca} - k_{abc}),$$

and hence that

$$\Gamma^a_{bc} = \tfrac{1}{2}h^{ad}(k_{cdb} + k_{bcd} - k_{dbc}),$$

where $h^{ab}h_{bc} = \delta^a_c$; that is, (h^{ab}) is the inverse of the matrix (h_{ab}). Conversely, if we define Γ^a_{bc} in this way, then we get

$$k_{cab} - \Gamma_{abc} - \Gamma_{bac}$$
$$= k_{cab} - \tfrac{1}{2}(k_{cab} + k_{bca} - k_{abc} + k_{cba} + k_{acb} - k_{bac})$$
$$= 0,$$

because $k_{abc} = k_{acb}$. $\square$

Note that

$$\Gamma^a_{bc} = \tfrac{1}{2}g^{ad}(\partial_c g_{db} + \partial_b g_{dc} - \partial_d g_{bc}),$$

evaluated at $x^a = 0$, where the g^{ab}s are the inverse or *contravariant* metric coefficients, defined by $g^{ab}g_{bc} = \delta^a_c$. The quantities Γ^a_{bc} are called the *Christoffel symbols*. We meet them again in the definition of the Levi-Civita connection.

Proposition 4.3

Let A be an event. Suppose that we have two coordinate systems x^a and $\tilde{x}^a$ such that $x^a = \tilde{x}^a = 0$ and $\partial_a g_{bc} = \tilde{\partial}_a \tilde{g}_{bc} = 0$ at A. Then there exist constants M^a_b such that $x^a = M^a_b \tilde{x}^b + O(3)$.

Here '$O(3)$' denotes third-order terms in x. The proposition says that the transformation is linear at A up to the second order in x; that is, the Taylor expansion about A of x^a in powers $\tilde{x}^a$ has no second-order terms.

Proof

We have to show that $\partial^2 x^a / \partial \tilde{x}^b \partial \tilde{x}^c = 0$ at A. Now at all events,

$$\tilde{g}_{ab} = g_{cd} \frac{\partial x^c}{\partial \tilde{x}^a} \frac{\partial x^d}{\partial \tilde{x}^b}.$$

Therefore

$$\tilde{\partial}_e \tilde{g}_{ab} = \frac{\partial x^f}{\partial \tilde{x}^e} \partial_f g_{cd} \frac{\partial x^c}{\partial \tilde{x}^a} \frac{\partial x^d}{\partial \tilde{x}^b} + g_{cd} \frac{\partial^2 x^c}{\partial \tilde{x}^a \partial \tilde{x}^e} \frac{\partial x^d}{\partial \tilde{x}^b} + g_{cd} \frac{\partial x^c}{\partial \tilde{x}^a} \frac{\partial^2 x^d}{\partial \tilde{x}^b \partial \tilde{x}^e}. \tag{4.6}$$

Note that the second two terms on the right-hand side differ by the interchange of a and b. Put

$$L_{abe} = g_{cd} \frac{\partial x^c}{\partial \tilde{x}^a} \frac{\partial^2 x^d}{\partial \tilde{x}^b \partial \tilde{x}^e}.$$

Then $L_{abe} = L_{aeb}$. Because the partial derivatives of g_{ab} and of $\tilde{g}_{ab}$ vanish at A, eqn (4.6) gives

$$\begin{aligned} L_{bae} + L_{abe} &= 0 \\ L_{eba} + L_{bea} &= 0 \\ L_{aeb} + L_{eab} &= 0. \end{aligned}$$

By adding the first and third, and subtracting the second, we obtain $L_{abe} = 0$. Hence

$$\frac{\partial^2 x^q}{\partial \tilde{x}^b \partial \tilde{x}^e} = \frac{\partial \tilde{x}^a}{\partial x^p} g^{pq} g_{cd} \frac{\partial x^c}{\partial \tilde{x}^a} \frac{\partial^2 x^d}{\partial \tilde{x}^b \partial \tilde{x}^e} = \frac{\partial \tilde{x}^a}{\partial x^p} g^{pq} L_{abe} = 0,$$

which completes the proof.   □

Proposition 4.4 (Existence of local inertial coordinates)

Let $g_{ab}(x)$ be a set of metric coefficients satisfying (MC1)–(MC4) and let A be an event. Then there exists a coordinate system x^a such that $x^a = 0$ at A and

$$\big(g_{ab}(x)\big) = \begin{pmatrix} 1 & 0 & 0 & 0 \\ 0 & -1 & 0 & 0 \\ 0 & 0 & -1 & 0 \\ 0 & 0 & 0 & -1 \end{pmatrix} + O(2)$$

as $x^a \to 0$. The system is unique up to coordinate transformations of the form

$$x^a = L^a{}_b \tilde{x}^b + O(3),$$

where $L = (L^a{}_b)$ is a Lorentz transformation matrix.

Proof

Choose an initial coordinate system such that $\partial_c g_{ab} = 0$ and $x^a = 0$ at A. Let h denote the matrix of metric coefficients at $x^a = 0$. Because h has signature $+ - - -$, we can find a matrix $J = (J^a_b)$ such that

$$J^t h J = \begin{pmatrix} 1 & 0 & 0 & 0 \\ 0 & -1 & 0 & 0 \\ 0 & 0 & -1 & 0 \\ 0 & 0 & 0 & -1 \end{pmatrix}.$$

Now make a linear coordinate change by replacing x^a by $J^a_b x^b$ to get the existence statement. The uniqueness statement follows from the previous proposition. □

The coordinates at A in the last proposition are interpreted as local inertial coordinates of an observer in free-fall at A. For special metrics we can reduce g_{ab} to the diagonal form $\mathrm{diag}\,(1, -1, -1, -1)$ everywhere. We show that this happens when the gravitational field vanishes. For a general metric, however, such a coordinate transformation does not exist. To summarize:

(1) A gravitational field is described by a general set of metric coefficients satisfying (MC1)–(MC4), which encode the temporal and spatial separation of nearby events.

(2) The local inertial coordinates set up by an observer in free-fall at an event A are the coordinates x^a such that $x^a = 0$ at A and

$$(g_{ab}) = \begin{pmatrix} 1 & 0 & 0 & 0 \\ 0 & -1 & 0 & 0 \\ 0 & 0 & -1 & 0 \\ 0 & 0 & 0 & -1 \end{pmatrix} + O(2)$$

as $x^a \to 0$. In local inertial coordinates, special relativity holds over small times and distances.

4.3 Particle Motion

In a local inertial coordinate system at an event A, $\partial_c g_{ab} = 0$ at A. The worldlines of massive free particles—particles in free-fall—satisfy

$$\ddot{x}^a = 0 \tag{4.7}$$

at A, where the dot is differentiation with respect to proper time τ. This equation determines their motion, but not in a very practical way because we have to use a different coordinate system at each event. To find the particle worldlines in a gravitational field, we need first to re-express (4.7) in a general coordinate system. To do this, we use the machinery of analytical dynamics, which is well suited to the purpose of writing down equations of motion in classical mechanics in general coordinate systems. Our strategy is to find a Lagrangian and to use a result from classical mechanics about the transformation of Lagrange's equations under change of coordinates.

Invariance of Lagrange's equations

The equations of motion of a classical dynamical system with time-independent Lagrangian $L(q_a, \dot{q}_a)$ are Lagrange's equations,

$$\frac{\mathrm{d}}{\mathrm{d}t}\left(\frac{\partial L}{\partial \dot{q}_a}\right) - \frac{\partial L}{\partial q_a} = 0\,,$$

where the q_as are generalized coordinates. The equations in a new coordinate system $\tilde{q}_a$ can be found by substituting

$$q_a = q_a(\tilde{q}), \qquad \dot{q}_a = \frac{\partial q_a}{\partial \tilde{q}_b}\dot{\tilde{q}}_b$$

into L and by writing down Lagrange's equations in the new coordinates. This is the sense in which Lagrange's equations are invariant under coordinate transformations.

The result has deep physical significance, but as a mathematical proposition, it is simply a statement about how a particular system of second-order differential equations changes when new dependent variables are substituted for the originals. If a system of ordinary differential equations for the functions $q_a(t)$ of a variable t can be written in the form of Lagrange's equations, then the transformed equations are of the same form, with the new Lagrangian found from the original by expressing q_a and $\dot{q}_a$ in terms of $\tilde{q}_a$ and $\dot{\tilde{q}}_a$.

So we can take the result out of its original physical context and use it to write the equations of motion of a freely falling particle in a general coordinate system. In the new context, we put the space–time coordinates x^a in the role of the q_as and the proper time τ in the role of time in classical mechanics. For the Lagrangian we take $L = \frac{1}{2}g_{ab}\dot{x}^a\dot{x}^b$, where the dot denotes differentiation with respect to proper time τ. The corresponding Lagrange equations are

$$\frac{\mathrm{d}}{\mathrm{d}\tau}\left(\frac{\partial L}{\partial \dot{x}^a}\right) - \frac{\partial L}{\partial x^a} = 0\,.$$

They are called the *geodesic equations*, and the solution curves in space–time are called *geodesics*.

Proposition 4.5

The geodesic equations are equivalent to

$$\ddot{x}^a + \Gamma^a_{bc}\dot{x}^b\dot{x}^c = 0\,,$$

where the Γ^a_{bc}s are the Christoffel symbols. The equations are invariant: that is, they take the same form in every coordinate system. In local inertial coordinates at an event, they reduce to $\ddot{x}^a = 0$ at the event.

Proof

To establish the first statement, we write out the geodesic equations explicitly. They are

$$\frac{\mathrm{d}}{\mathrm{d}\tau}\left(g_{ab}\dot{x}^b\right) - \tfrac{1}{2}(\partial_a g_{bc})\dot{x}^b\dot{x}^c = 0.$$

That is,

$$g_{db}\ddot{x}^b + \tfrac{1}{2}\dot{x}^b\dot{x}^c(2\partial_c g_{db} - \partial_d g_{bc}) = 0\,,$$

by changing a to d and by using $\dot{g}_{bc} = \dot{x}^c\partial_c g_{ab}$. By multiplying by the inverse metric g^{ad} and by using the symmetry under interchange of the dummy indices b and c, we can rewrite this as

$$\ddot{x}^a + \tfrac{1}{2}\dot{x}^b\dot{x}^c g^{ad}(\partial_b g_{dc} + \partial_c g_{bd} - \partial_d g_{bc}) = 0\,.$$

In other words,

$$\ddot{x}^a + \Gamma^a_{bc}\dot{x}^b\dot{x}^c = 0\,,$$

where

$$\Gamma^a_{bc} = \tfrac{1}{2}g^{ad}(\partial_b g_{dc} + \partial_c g_{bd} - \partial_d g_{bc}). \tag{4.8}$$

These are the *Christoffel symbols* or *connection coefficients*, which have already appeared on page 47.

The invariance of the equations follows from the invariance of L. From (MC3),

$$\tilde{g}_{ab}\dot{\tilde{x}}^a\dot{\tilde{x}}^b = \tilde{g}_{ab}\frac{\partial\tilde{x}^a}{\partial x^c}\frac{\partial\tilde{x}^b}{\partial x^d}\dot{x}^c\dot{x}^d = g_{cd}\dot{x}^c\dot{x}^d\,.$$

Thus the geodesic equations take the same form in every coordinate system; and in local inertial coordinates at an event they reduce to the equations of motion of a free-falling particle. They hold in a special coordinate system at each event; therefore they hold in every coordinate system at every event and so determine the motion of the particle in any coordinate system.

Because the Christoffel symbols vanish at an event A in local inertial coordinates at A, the equations reduce to $\ddot{x}^a = 0$ in these coordinates at the event. $\qquad\square$

The motion of a particle in free-fall is therefore given by the geodesic equations in local inertial coordinates at an event, and hence in any coordinates. We are led to the following.

The geodesic hypothesis

The worldlines of particles in free-fall satisfy the geodesic equations, with τ the proper time.

It follows from the geodesic equations that L is constant. This can be shown by direct calculation, or by appealing to the fact that L is a homogeneous quadratic in the $\dot{x}^a$s and has no explicit dependence on proper time.[1] In fact, on a particle worldline parametrized by proper time,

$$L = \tfrac{1}{2}g_{ab}\dot{x}^a\dot{x}^b = \tfrac{1}{2}\,.$$

Example 4.6

In Minkowski space in spherical polar coordinates,

$$L = \tfrac{1}{2}(\dot{t}^2 - \dot{r}^2 - r^2\dot{\theta}^2 - r^2\sin^2\theta\dot{\varphi}^2)\,.$$

The geodesic equations are

$$\ddot{t} = 0 \qquad\qquad \ddot{\theta} + 2r^{-1}\dot{r}\dot{\theta} - \sin\theta\cos\theta\dot{\varphi}^2 = 0$$
$$\ddot{r} - r\dot{\theta}^2 - r\sin^2\theta\dot{\varphi}^2 = 0, \qquad \ddot{\varphi} + 2r^{-1}\dot{r}\dot{\varphi} + 2\cot\theta\dot{\varphi} = 0\,.$$

We can read off from these that, for example, $\Gamma^3_{13} = 1/r$, with coordinates ordered so that $x^0 = t$, $x^1 = r$, $x^2 = \theta$, $x^3 = \varphi$.

4.4 Null Geodesics

By the same reasoning, the worldline of a photon is also given by the geodesic equations,

$$\frac{\mathrm{d}}{\mathrm{d}\tau}\left(\frac{\partial L}{\partial \dot{x}^a}\right) - \frac{\partial L}{\partial x^a} = 0\,,$$

---

[1] In analytical dynamics, the Hamiltonian is conserved whenever the Lagrangian has no explicit time dependence; and if the Lagrangian is a homogeneous quadratic, then it is the same as the Hamiltonian. Again these statements can be taken out of their original physical context and interpreted as propositions concerning a Lagrangian system of ordinary differential equations.

where τ is now an affine parameter. In this case,

$$L = \tfrac{1}{2}g_{ab}\dot{x}^a\dot{x}^b = 0$$

because $ds^2 = g_{ab}\,dx^a\,dx^b = 0$ for two nearby events on the worldline of a photon.

Geodesics with $g_{ab}\dot{x}^a\dot{x}^b > 0$ are said to be *timelike*; those with $g_{ab}\dot{x}^a\dot{x}^b = 0$ are said to be *null*. So photon worldlines are null geodesics and massive particle worldlines are timelike geodesics.

4.5 Transformation of the Christoffel Symbols

The Christoffel symbols are defined by (4.8). They determine the worldlines of free particles through the geodesic equations, and so contain the same information as the 'acceleration due to gravity' in Newtonian theory. They vanish at the origin in local inertial coordinates, as one would expect: local inertial coordinates are the coordinates set up by an observer in free-fall at an event. In the observer's frame, the 'acceleration due to gravity' is zero.

How do the Christoffel symbols transform when we change coordinates from one general system x^a to another $\tilde{x}^a$? In the new coordinates,

$$\tilde{\Gamma}^a_{bc} = \tfrac{1}{2}\tilde{g}^{ad}(\tilde{\partial}_b\tilde{g}_{dc} + \tilde{\partial}_c\tilde{g}_{ba} - \tilde{\partial}_a\tilde{g}_{bc}).$$

We could determine the relationship between Γ^a_{bc} and $\tilde{\Gamma}^a_{bc}$ by direct substitution. But the calculation is unnecessarily complicated. Instead, we use the fact that the geodesic equations

$$\ddot{x}^a + \Gamma^a_{bc}\dot{x}^b\dot{x}^c = 0 \tag{4.9}$$

transform to

$$\ddot{\tilde{x}}^a + \tilde{\Gamma}^a_{bc}\dot{\tilde{x}}^b\dot{\tilde{x}}^c = 0$$

because the Lagrangian from which they are derived is invariant. Substitute

$$\dot{\tilde{x}}^a = \frac{\partial\tilde{x}^a}{\partial x^d}\dot{x}^d$$

into the second equation to get

$$
\begin{aligned}
0 &= \frac{\partial\tilde{x}^a}{\partial x^d}\ddot{x}^d + \frac{\partial^2\tilde{x}^a}{\partial x^d\partial x^e}\dot{x}^d\dot{x}^e + \tilde{\Gamma}^a_{ef}\frac{\partial\tilde{x}^e}{\partial x^b}\frac{\partial\tilde{x}^f}{\partial x^c}\dot{x}^b\dot{x}^c \\
\Rightarrow\quad 0 &= \ddot{x}^p + \frac{\partial x^p}{\partial\tilde{x}^d}\left[\tilde{\Gamma}^d_{ef}\frac{\partial\tilde{x}^e}{\partial x^b}\frac{\partial\tilde{x}^f}{\partial x^c} + \frac{\partial^2\tilde{x}^d}{\partial x^b\partial x^c}\right]\dot{x}^b\dot{x}^c, \tag{4.10}
\end{aligned}
$$

with the second line following from the first by multiplying by $\partial x^p / \partial \tilde{x}^a$ and summing over a. Hence because $\Gamma^a_{bc} = \Gamma^a_{cb}$, and because (4.10) and (4.9) are equivalent for all choices of free particle worldline,

$$\Gamma^a_{bc} = \frac{\partial x^a}{\partial \tilde{x}^d} \tilde{\Gamma}^d_{ef} \frac{\partial \tilde{x}^e}{\partial x^b} \frac{\partial \tilde{x}^f}{\partial x^c} + \frac{\partial x^a}{\partial \tilde{x}^d} \frac{\partial^2 \tilde{x}^d}{\partial x^b \partial x^c} \, .$$

The first term on the right could have been anticipated: it is simply the tensor transformation rule. The second involves the second derivative of the new coordinates with respect to the old. Thus it measures, in some sense, the acceleration of the new coordinates relative to the old. It should also have been anticipated, because it mirrors the acceleration term in the transformation of $\boldsymbol{g}$ when one switches to an accelerating frame in Newtonian theory.

4.6 Manifolds

We now have one half of general relativity: we know how gravity affects matter. The gravitational field is encoded in the metric coefficients g_{ab}, and the motion of a freely falling particle is governed by the geodesic equations. Gravity is not a field, like the electromagnetic field, but is part of the structure of space–time.

So what sort of object is the space–time of general relativity? In local inertial coordinates, it looks in a small region like Minkowski space; but when we extend the coordinates over a larger region, the light cones are not fixed: they vary from event to event. We have the analogy with the relationship between a curved surface and a flat plane. The local geometry is the same: we can map a small part of the earth's surface onto a page in an atlas with a constant scale; but a map of a large region will introduce distortion. Analogously, a small region of space–time can be mapped onto Minkowski space by using local inertial coordinates, but as we extend the coordinates to a larger region, the identification breaks down. The geodesics in space–time are not mapped onto straight lines.

A space–time in general relativity and a surface in space are examples of *manifolds*, that is, spaces whose points can be labelled by coordinates. In relativity, events are labelled by four space–time coordinates x^a; on a surface, we use two parameters, such as latitude and longitude on the sphere, to label the individual points. In neither case is there a natural choice for the coordinates, and it may be impossible to use a single coordinate system to cover the whole space. Longitude, for example, is not uniquely defined at the North and South poles. So the definition of a manifold captures the idea that the coordinate systems are local, and ties down the permitted transformations between local coordinates. There are many possibilities, but we only allow smooth, that is to say infinitely differentiable, transformations. Our manifolds are therefore of class C^∞.

Definition 4.7

An n-dimensional manifold is

(a) A connected Hausdorff topological space M, together with

(b) A collection of *charts* or *coordinate patches* (U, x^a), where $U \subset M$ is an open set and the x^as are n functions $x^a : U \to \mathbb{R}$, such that the map

$$\boldsymbol{x} : U \to \mathbb{R}^n : m \mapsto \big(x^0(m), x^1(m), \ldots, x^{n-1}(m)\big)$$

is a homeomorphism from U to an open subset $V \subset \mathbb{R}^n$.

Two conditions must hold: (i) every point of M must lie in a coordinate patch; and (ii) if (U, x^a) and $(\tilde{U}, \tilde{x}^a)$ are charts, then the $\tilde{x}^a$s can be expressed as functions of the x^as on the intersection. We require that

$$\boldsymbol{x}(U \cap \tilde{U}) \to \tilde{\boldsymbol{x}}(U \cap \tilde{U}) : (x^a) \mapsto (\tilde{x}^a)$$

should be infinitely differentiable and one-to-one, with

$$\det \left[\frac{\partial x^a}{\partial \tilde{x}^b} \right] \neq 0.$$

The topological condition on M is required to rule out pathological behaviour. In fact further technical conditions, such as 'paracompactness', are needed to get sensible models of space–time. We should also specify completeness for the atlas (the set of charts). We do not dwell on such matters here because they play no part in the elementary development of the theory. Topological language is needed only to give meaning to the term 'local coordinates': local coordinates label the points of open sets of M, and the transformations between local coordinate systems are smooth and invertible.

A surface is a two-dimensional manifold; space–time is a four-dimensional manifold. Both have an additional structure called a *metric*. On a surface, the metric determines the geometry: it gives the distance between nearby points. If the surface is defined parametrically by giving the position $\boldsymbol{r}$ of a general point as a function $\boldsymbol{r}(u, v)$ of two parameters, then the distance $\mathrm{d}s$ between the nearby points (u, v) and $(u + \mathrm{d}u, v + \mathrm{d}v)$ is determined by

$$
\begin{aligned}
\mathrm{d}s^2 &= \boldsymbol{r}_u.\boldsymbol{r}_u \, \mathrm{d}u^2 + 2\boldsymbol{r}_u.\boldsymbol{r}_v \, \mathrm{d}u\mathrm{d}v + \boldsymbol{r}_v.\boldsymbol{r}_v \, \mathrm{d}v^2 \\
&= E \, \mathrm{d}u^2 + 2F \, \mathrm{d}u\mathrm{d}v + G \, \mathrm{d}v^2 \,,
\end{aligned}
\tag{4.11}
$$

where $E = \boldsymbol{r}_u.\boldsymbol{r}_u$ and so on. This is the *first fundamental form*. Like $\mathrm{d}s^2$ in space–time, it is a quadratic form in the coordinate displacement. It measures the separation between two nearby points on the surface. The coefficients E, F, G are functions of the 'coordinates' u, v, like the metric coefficients g_{ab} in space–time. We note the following.

(1) In general, the metric cannot be reduced to the flat form $\mathrm{d}u^2 + \mathrm{d}v^2$ by changing the parameters. This is only possible if the surface has no *intrinsic* or *Gaussian curvature*. We establish this in §4.8.

(2) The surface may have nontrivial topology, in which case the same parameters cannot be used over the entire surface. In general relativity, similarly, we must allow for space–time to have a nontrivial topology. This is important in the model space–times used in cosmology.

An expression such as (4.11) is manageable when there are only two coordinates and three metric coefficients. In higher dimensions, one needs a more compact and efficient way of representing the metric and doing calculations involving the metric coefficients. This is provided by *tensor calculus*, in which the space–time metric, and other physical quantities, are represented by tensors. We look at the definitions only in the four dimensions of space–time, although the extension to the general setting of an n-dimensional manifold is obvious.

4.7 Vectors and Tensors

The various physical objects in space–time are represented by scalars—functions on space–time—or by vectors or tensors, which are objects with components that transform in simple ways under change of coordinates. The definitions are the same as in special relativity, except that the coordinate changes are now general.

Definition 4.8

A tensor T of type (p, q) is an object that assigns a set of components $T^{a...b}{}_{c...d}$ (p upper indices, q lower indices) to each local coordinate system, with the transformation rule under change of coordinates

$$T^{a...b}{}_{c...d} = \frac{\partial x^a}{\partial \tilde{x}^e} \cdots \frac{\partial x^b}{\partial \tilde{x}^f} \frac{\partial \tilde{x}^h}{\partial x^c} \cdots \frac{\partial \tilde{x}^k}{\partial x^d} \tilde{T}^{e...f}{}_{h...k}.$$

A tensor can be defined at a single event, or along a curve, or on the whole of space–time, in which case the components are functions of the coordinates and we call T a *tensor field*. If $q = 0$ then T is a *contravariant* tensor, if $p = 0$, it is a *covariant* tensor. A tensor of type $(1, 0)$ is a *four-vector* or simply a *vector*.

An object that behaves as a tensor under change of local inertial coordinates at an event determines a tensor at the event under general coordinate transformations. We frequently fail to distinguish between a tensor and its components, and allow ourselves the usage 'a tensor $T^{a...b}{}_{c...d}$' or 'a vector V^a'.

Note that because

$$\frac{\partial x^a}{\partial \tilde{x}^e} \frac{\partial \tilde{x}^e}{\partial x^b} = \delta_b^a \, , \tag{4.12}$$

one could equally well write the transformation law with all the tilded (˜) and untilded quantities interchanged.

Example 4.9

The metric g_{ab} is a tensor field of type $(0, 2)$. It has the transformation law

$$g_{ab} = \tilde{g}_{cd} \frac{\partial \tilde{x}^c}{\partial x^a} \frac{\partial \tilde{x}^d}{\partial x^b} \, . \tag{4.13}$$

Example 4.10

The contravariant metric has components g^{ab}, where (g^{ab}) is the inverse matrix to (g_{ab}). That is, $g^{ab} g_{bc} = \delta_c^a$. It is a tensor of type $(2, 0)$. This is proved from (4.13) by the following steps, which are well worth following carefully because they illustrate some basic techniques of index manipulation. The proof makes several uses of (4.12). First, multiply both sides of (4.13) by

$$\frac{\partial \tilde{x}^b}{\partial x^e}$$

and sum over b. The result is

$$\tilde{g}_{ab} \frac{\partial \tilde{x}^b}{\partial x^e} = g_{ce} \frac{\partial x^c}{\partial \tilde{x}^a} \, .$$

Now multiply by $\tilde{g}^{af} g^{eh}$ and sum over a, e to get

$$\frac{\partial \tilde{x}^f}{\partial x^e} g^{eh} = \tilde{g}^{af} \frac{\partial x^h}{\partial \tilde{x}^a} \, .$$

Finally multiply by

$$\frac{\partial x^k}{\partial \tilde{x}^f}$$

and sum over f to get

$$g^{kh} = \tilde{g}^{af} \frac{\partial x^h}{\partial \tilde{x}^a} \frac{\partial x^k}{\partial \tilde{x}^f} \, .$$

Example 4.11

The gradient $\partial_a f$ of a scalar function is a covector field, a tensor of type $(0, 1)$.

Example 4.12

If $x^a = x^a(\tau)$ is the worldline of a particle in general motion, parametrized by a parameter τ, then

$$V^a = \frac{\mathrm{d}x^a}{\mathrm{d}\tau}$$

is a four-vector field along the worldline. If

$$g_{ab} V^a V^b = g_{ab} \dot{x}^a \dot{x}^b = 1 \, ,$$

then V is called the *four-velocity* and τ is called the *proper time*. This extends the definition of proper time from motion in free-fall. When

$$g_{ab} V^a V^b = 1 \, ,$$

the increment in τ between the events on the worldline with coordinates x^a and $x^a + \mathrm{d}x^a$ is

$$\mathrm{d}\tau = \sqrt{g_{ab} \, \mathrm{d}x^a \, \mathrm{d}x^b} \, .$$

So by the interpretation of the metric, $\mathrm{d}\tau$ is the time between the two events measured in a local inertial frame in which they happen at the same place. Proper time therefore has the same meaning as in special relativity. We extend the *clock hypothesis* to the general setting by postulating that proper time is the time measured by a clock of standard construction travelling with the particle. As in special relativity, the mechanism of the clock must be insensitive to the acceleration of the particle (a pendulum clock will not do).

We can carry out all the operations on tensors in exactly the same way as in special relativity, with the exception of differentiation. Partial differentiation with respect to the coordinates no longer gives a tensor because the components $\partial_a T^{b\cdots}{}_{d\ldots}$ do not obey the tensor transformation law under nonlinear coordinate changes. Indices are raised and lowered by contracting with g^{ab} and g_{ab}, although this now involves more than just changing the signs of a few components. For example, if $T^{ab}{}_c$ is a tensor of type $(2, 1)$, then the contraction $T^{ab}{}_b$ is a tensor of type $(1, 0)$ (one free upper index a). If α_a is a covector, then $g^{ab}\alpha_c$ is a tensor of type $(2, 1)$ and its contraction

$$\alpha^a = g^{ab}\alpha_b$$

is a vector. This is the operation of *raising the index*. One similarly lowers indices, for example, by putting $X_a = g_{ab}X^b$. Raising followed by lowering returns to the starting point because

$$g_{ab}g^{bc} = \delta_a^c \,.$$

The exceptional operation, differentiation, is more subtle in a general space–time. We come back to it in the next chapter.

4.8 The Geometry of Surfaces*

Much of the general theory of relativity can be illuminated by exploring the analogy between the structure of space–time and the more familiar and more easily visualized geometry of a surface. This section summarizes the theory of surfaces in a way that may help to draw out the analogy. It is not essential to the following chapters, but we refer back to it from time to time to draw attention to the analogies.

The metric tensor on a surface determines the distance between nearby points. Its components

$$E = \boldsymbol{r}_u \cdot \boldsymbol{r}_u, \qquad F = \boldsymbol{r}_u \cdot \boldsymbol{r}_v, \qquad G = \boldsymbol{r}_v \cdot \boldsymbol{r}_v$$

can be read off from the the first fundamental form (4.11). Just as the metric coefficients in space–time, they transform as the components of a tensor of type $(2,0)$ under change of parametrization. The same argument as in §4.2 establishes that at any point p on the surface, it is possible to choose the parameters u, v so that $u = v = 0$ at p and

$$E = 1 + O(w^2), \qquad F = O(w^2), \qquad \text{and} \qquad G = 1 + O(w^2), \qquad (4.14)$$

as $u, v \to 0$, where $w = \sqrt{u^2 + v^2}$. We define the *Gaussian curvature* at p by

$$\kappa(p) = -\tfrac{1}{2}(E_{vv} + G_{uu} - 2F_{uv})$$

in this special parametrization (the subscripts denote partial derivatives). Of course the special parametrization in which (4.14) holds is not unique. So to establish that the definition is a good one, we need to show that the value of $\kappa(p)$ is independent of the choice made.

This is done by deriving another formula for $\kappa(p)$. At each point of the surface in a neighbourhood of p, choose two orthogonal unit vectors $\boldsymbol{a}$, $\boldsymbol{b}$ tangent to the surface, so that $\boldsymbol{a}$, $\boldsymbol{b}$, and the unit normal $\boldsymbol{n}$ to the surface make up a right-handed orthonormal triad (Figure 4.2). Given a curve $\boldsymbol{r} = \boldsymbol{r}(t)$ on the

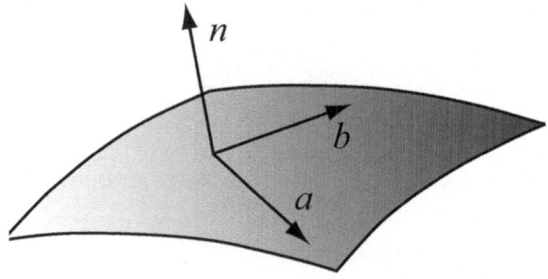

Figure 4.2 The triad $\boldsymbol{a}$, $\boldsymbol{b}$, $\boldsymbol{n}$

surface, consider the quantity defined along the curve by

$$\boldsymbol{a} \cdot \dot{\boldsymbol{b}} = -\boldsymbol{b} \cdot \dot{\boldsymbol{a}} = \boldsymbol{a} \cdot \left((\dot{\boldsymbol{r}} \cdot \boldsymbol{\nabla}) \boldsymbol{b} \right),$$

where $\boldsymbol{\nabla}$ is the three-dimensional gradient and the dot denotes differentiation with respect to t. This is linear in the tangent vector $\dot{\boldsymbol{r}}$. So there is a vector $\boldsymbol{\omega}$ tangent to the surface at each point such that

$$\boldsymbol{a} \cdot \dot{\boldsymbol{b}} = \dot{\boldsymbol{r}} \cdot \boldsymbol{\omega}$$

for any curve on the surface. It depends, of course, on the choice of $\boldsymbol{a}$, $\boldsymbol{b}$. If we make a rotation at each point and replace $\boldsymbol{a}$ and $\boldsymbol{b}$ by

$$\tilde{\boldsymbol{a}} = \cos\theta \, \boldsymbol{a} + \sin\theta \, \boldsymbol{b} \qquad \tilde{\boldsymbol{b}} = -\sin\theta \, \boldsymbol{a} + \cos\theta \, \boldsymbol{b},$$

where θ is a function of u, v, then $\boldsymbol{\omega}$ is replaced by $\boldsymbol{\omega} - \boldsymbol{\nabla}\theta$.

Proposition 4.13

$\kappa(p) = \boldsymbol{n} \cdot \operatorname{curl}\boldsymbol{\omega}$, evaluated at p.

Proof

Note, first, that $\boldsymbol{n} \cdot \operatorname{curl}\boldsymbol{\omega}$ is well defined because it involves only derivatives of $\boldsymbol{\omega}$ tangent to the surface. Equally it is independent of the choice of $\boldsymbol{a},\boldsymbol{b}$, because the curl of a gradient vanishes. In fact, if we use the subscripts $i, j, k, \dots$ to label Cartesian coordinates on $\mathbb{R}^3$, then

$$\boldsymbol{n} \cdot \operatorname{curl}\boldsymbol{\omega} = \epsilon_{ijk} n_i \partial_j \omega_k = \epsilon_{ijk} n_i \partial_j (a_l \partial_k b_l) = \epsilon_{ijk} n_i (\partial_j a_l)(\partial_k b_l).$$

But $\boldsymbol{a}$, $\boldsymbol{b}$, $\boldsymbol{n}$ form a right-handed triad. So the last expression is

$$(a_j \partial_j a_l)(b_k \partial_k b_p) - (b_j \partial_j a_l)(a_k \partial_k b_p)\,,$$

in which the Cartesian components of $\boldsymbol{a}$ and $\boldsymbol{b}$ are differentiated only along $\boldsymbol{a}$ and $\boldsymbol{b}$, which are tangent to the surface.

We can choose $\boldsymbol{a}$ and $\boldsymbol{b}$ so that, with the special choice of parameters,

$$\boldsymbol{a} = \boldsymbol{r}_u + O(w^2), \qquad \boldsymbol{b} = \boldsymbol{r}_v + O(w^2)$$

as $u, v \to 0$. We then have

$$
\begin{aligned}
\boldsymbol{n} \cdot \operatorname{curl} \boldsymbol{\omega}(p) &= (a_j \partial_j a_l)(b_k \partial_k b_p) - (b_j \partial_j a_l)(a_k \partial_k b_p) \\
&= \boldsymbol{r}_{uu} \cdot \boldsymbol{r}_{vv} - \boldsymbol{r}_{uv} \cdot \boldsymbol{r}_{uv}\,,
\end{aligned}
$$

evaluated at $u = v = 0$. At p, however,

$$E_u = 2\boldsymbol{r}_u \cdot \boldsymbol{r}_{uu} = 0, \quad E_v = 2\boldsymbol{r}_u \cdot \boldsymbol{r}_{uv} = 0, \quad F_u = \boldsymbol{r}_v \cdot \boldsymbol{r}_{uu} + \boldsymbol{r}_u \cdot \boldsymbol{r}_{uv} = 0\,.$$

From this and similar expressions for G_u, G_v, and F_v, we deduce that $\boldsymbol{r}_{uu}$, $\boldsymbol{r}_{uv}$, and $\boldsymbol{r}_{vv}$ are orthogonal to $\boldsymbol{r}_u$ and to $\boldsymbol{r}_v$ at p. By differentiating twice the defining equations of E, F, G with respect to u, v, we deduce that at p,

$$E_{uu} + G_{vv} - 2F_{uv} = 2(\boldsymbol{r}_{uv} \cdot \boldsymbol{r}_{uv} - \boldsymbol{r}_{uu} \cdot \boldsymbol{r}_{vv})\,,$$

which completes the proof. $\qquad\square$

Because $\boldsymbol{n} \cdot \operatorname{curl} \boldsymbol{\omega}$ does not depend on the choice of $\boldsymbol{a}, \boldsymbol{b}$, the value of $\kappa(p)$ does not depend on the choice of the special parameters u, v. So the Gaussian curvature is a well-defined function on the surface. If it does not vanish, then it is impossible to reduce the first fundamental form to the planar metric $\mathrm{d}u^2 + \mathrm{d}v^2$ throughout the u,v coordinate patch. The fact that κ can be computed from the first fundamental form alone is Gauss's *theorema egregium*.

The Gaussian curvature measures the extent to which the geometry of the surface differs from that of the flat plane. One of the most direct ways in which it can be interpreted is in terms of the excess of the sum of the angles of a geodesic triangle over π. A geodesic on the surface is the closest that a curve on the surface can come to being a 'straight line' without leaving the surface. It is the path followed by a particle constrained to move on the surface by a 'normal reaction' (in the direction of $\boldsymbol{n}$), in the absence of other forces. It is also the curve that minimizes distance between two nearby points.

In close analogy to the space–time theory, the geodesics are generated by the Lagrangian

$$L = \tfrac{1}{2}(E\dot{u}^2 + 2F\dot{u}\dot{v} + G\dot{v}^2)$$

where u, v are general coordinates. When $L = \frac{1}{2}$, the parameter on the geodesic is the arclength s. The connection with the motion of a particle comes from identifying L with the kinetic energy $\frac{1}{2}\dot{\boldsymbol{r}} \cdot \dot{\boldsymbol{r}}$ of a unit mass particle constrained to move on the surface, with unit speed.

With arclength as parameter, the tangent $\boldsymbol{t} = \dot{\boldsymbol{r}}$ to a geodesic is a unit vector. Its derivative $\dot{\boldsymbol{t}}$ is given in a general parametrization of the surface by

$$\dot{\boldsymbol{t}} = \boldsymbol{r}_u \ddot{u} + \boldsymbol{r}_v \ddot{v} + \boldsymbol{r}_{uu} \dot{u}^2 + 2\boldsymbol{r}_{uv} \dot{u}\dot{v} + \boldsymbol{r}_{vv} \dot{v}^2 .$$

In the special coordinates at a point p, the geodesic equations reduce to $\ddot{u} = \ddot{v} = 0$ at p, and the second derivatives of $\boldsymbol{r}$ are orthogonal to the surface at p. We deduce that $\dot{\boldsymbol{t}}$ is orthogonal to the surface at p, and by the same argument, at every point of the geodesic. Thus the acceleration of a geodesic is everywhere in the direction on $\boldsymbol{n}$, as consideration of the equation of motion of the corresponding particle implies. The direction of $\boldsymbol{t}$ changes only as much as is necessary to follow the surface.

If we choose orthogonal vectors $\boldsymbol{a}$ and $\boldsymbol{b}$ as before, then

$$\boldsymbol{t} = \cos\theta\, \boldsymbol{a} + \sin\theta\, \boldsymbol{b} ,$$

for some function $\theta(s)$. Because $\boldsymbol{a} \cdot \dot{\boldsymbol{a}} = \boldsymbol{b} \cdot \dot{\boldsymbol{b}} = 0$ and $\boldsymbol{b} \cdot \dot{\boldsymbol{a}} = -\boldsymbol{a} \cdot \dot{\boldsymbol{b}}$, we have,

$$0 = (-\sin\theta\, \boldsymbol{a} + \cos\theta\, \boldsymbol{b}) \cdot \boldsymbol{t} = \dot{\theta} - \boldsymbol{a} \cdot \dot{\boldsymbol{b}} = \dot{\theta} - \dot{\boldsymbol{r}} \cdot \boldsymbol{\omega} . \qquad (4.15)$$

Now consider a triangle A, B, C on the surface, the sides of which are geodesics (Figure 4.3). We make the arclength increase along the three geodesics

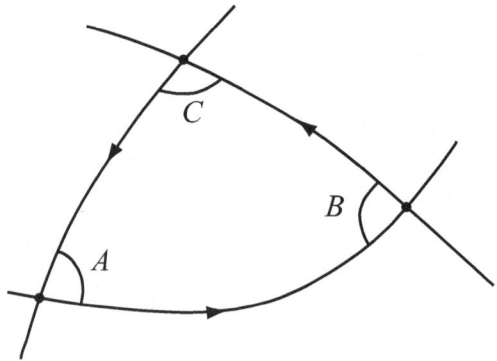

Figure 4.3 A geodesic triangle

from A to B, from B to C, and from C to A, and express the dependence of

θ on the three sides by $\theta_3(s)$, $\theta_1(s)$, and $\theta_2(s)$, respectively. We assume that the triangle is contained in the region in which the triad is defined. Then by integrating (4.15) around the triangle, we find

$$\theta_3(B) - \theta_3(A) + \theta_1(C) - \theta_1(B) + \theta_2(A) - \theta_2(C) = \oint \boldsymbol{\omega} \cdot \mathbf{d}\boldsymbol{r}\,.$$

By using Proposition 4.13 and by applying Stokes' theorem,

$$\oint \boldsymbol{\omega} \cdot \mathbf{d}\boldsymbol{r} = \int \mathrm{curl}\,\boldsymbol{\omega} \cdot \mathbf{d}\boldsymbol{S} = \int \kappa\,\mathrm{d}S\,,$$

where the second two integrals are over the interior of the triangle, and we have assumed that the interior of the triangle is simply connected. But

$$\theta_2(A) - \theta_3(A)$$

is the angle that $\boldsymbol{t}$ turns through at A in passing from the geodesic CA to the geodesic AB. The conclusion is the Gauss–Bonnet theorem.

Theorem 4.14 (Gauss–Bonnet)

The sum of the interior angles A, B, C of a small geodesic triangle is

$$A + B + C = \pi + \int \kappa\,\mathrm{d}S\,,$$

where the integral is over the interior of the triangle.

A rather more suggestive way to state the theorem, at least in the context of relativity, is in terms of the velocities of particles moving along geodesics on the surface, with no friction. Suppose that O travels from A to B, Q travels from B to C, and P travels directly from A to C. Let θ_A denote the angle between the velocities of P and O at A, θ_B the angle between the velocities of O and Q at B, and θ_C the angle between the velocities of Q and P at C (all assumed acute). Then

$$\theta_A - \theta_B + \theta_C = \int \kappa\,\mathrm{d}S\,.$$

In the plane, the left-hand side would be zero.

4.9 Summary of the Mathematical Formulation

Space–time is a four-dimensional manifold M with a metric tensor g_{ab}, which is a symmetric tensor of type (0,2) with signature $+---$. The points of M are the *events*. If $x^a = x^a(u)$ is the worldline of a particle, where u is a parameter, then

$$\tau = \int \sqrt{g_{ab} \frac{\mathrm{d}x^a}{\mathrm{d}u} \frac{\mathrm{d}x^b}{\mathrm{d}u}} \, \mathrm{d}u$$

is the proper time along the worldline, that is, the time measured by a clock carried by the particle. This is the *clock hypothesis*. The four-vector V with components $V^a = \mathrm{d}x^a/\mathrm{d}\tau$ is the particle's four-velocity.

The metric determines the behaviour of free particles via the geodesic hypothesis

$$\frac{\mathrm{d}^2 x^a}{\mathrm{d}\tau^2} + \Gamma^a_{bc} \frac{\mathrm{d}x^b}{\mathrm{d}\tau} \frac{\mathrm{d}x^c}{\mathrm{d}\tau} = 0 \,,$$

where τ is proper time for a particle with mass, or an affine parameter in the case of a photon.

If A is an event, then there exists a local coordinate system such that $x^a = 0$ at A and

$$\bigl(g_{ab}(x)\bigr) = \begin{pmatrix} 1 & 0 & 0 & 0 \\ 0 & -1 & 0 & 0 \\ 0 & 0 & -1 & 0 \\ 0 & 0 & 0 & -1 \end{pmatrix} + O(2)$$

as $x^a \to 0$. In these coordinates, $\Gamma^a_{bc} = 0$ at the origin (the event A). Such a coordinate system is interpreted as the local inertial coordinate system set up by an observer in free-fall at A. We identify four-vectors and tensors at A with vectors and tensors in special relativity by taking their components in local inertial coordinates.

The metric determines an inner product $g(X, Y) = X_a Y^a$ on the space of four-vectors at an event with signature $+ - --$. It is symmetric and nondegenerate, but not positive definite. As in special relativity, we say that X is timelike if $X^a X_a > 0$, null if $X^a X_a = 0$, and spacelike if $X^a X_a < 0$.

EXERCISES

4.1. Show that if x^a and $\tilde{x}^a$ are coordinate systems, then

$$\frac{\partial \tilde{x}^a}{\partial x^p} \frac{\partial^2 x^p}{\partial \tilde{x}^b \partial \tilde{x}^c} = -\frac{\partial x^q}{\partial \tilde{x}^b} \frac{\partial x^r}{\partial \tilde{x}^c} \frac{\partial^2 \tilde{x}^a}{\partial x^q \partial x^r} \,.$$

4.2. Show that if X and Y are vector fields on a manifold, then so is

$$Z^a = X^b \partial_b Y^a - Y^b \partial_b X^a \,.$$

That is, show that the Z^as transform correctly under change of co-ordinates.

4.3. Let $x^a(\tau)$ be a solution curve of the Lagrange equations of the Lagrangian $L = \frac{1}{2} g_{ab} \dot{x}^a \dot{x}^b$. Show from the Lagrange equations without assuming in advance that τ is proper time that

$$\frac{\mathrm{d}}{\mathrm{d}\tau} (g_{ab} V^a V^b) = 0.$$

How could you have deduced this directly from the Lagrangian?

4.4. Einstein proposed the following metric as a model for a closed static universe
$$\mathrm{d}s^2 = \mathrm{d}t^2 - \mathrm{d}r^2 - \sin^2 r (\mathrm{d}\theta^2 + \sin^2 \theta \mathrm{d}\varphi^2) \,.$$

Find the geodesic equations of the metric from Lagrange's equations and hence write down the Christoffel symbols (take $x^0 = t$, $x^1 = r$, $x^2 = \theta$, $x^3 = \varphi$). Show that there are geodesics on which r and θ are constant and equal to $\pi/2$.

4.5. The Einstein static universe is mapped into the five-dimensional space–time with metric

$$\mathrm{d}S^2 = \mathrm{d}T^2 - \mathrm{d}X^2 - \mathrm{d}Y^2 - \mathrm{d}Z^2 - \mathrm{d}W^2$$

by $T = t$, $X = \sin r \sin \theta \sin \varphi$, $Y = \sin r \sin \theta \cos \varphi$, $Z = \sin r \cos \theta$, and $W = \cos r$. Show that $\mathrm{d}s^2 = \mathrm{d}S^2$.

Show that the image is (almost all of) $\{X^2 + Y^2 + Z^2 + W^2 = 1\}$. Deduce that, as a topological space, the Einstein universe is the product of $\mathbb{R}$ and the three-dimensional sphere $X^2 + Y^2 + Z^2 + W^2 = 1$ in $\mathbb{R}^4$. What portion is covered by the chart t, r, θ, φ? Describe the geodesic curves on the image.

5
Tensor Calculus

We have seen that the space–time of general relativity is a four-dimensional manifold and that gravity is encoded in the metric tensor. It manifests itself in the relative acceleration of local inertial frames, and thus in variations in the metric from event to event.

Our next task is to understand how matter generates gravity; that is, how to relate the variations in the metric to the distribution of matter in space–time. To do this, we must know how to differentiate vectors and tensors. In Minkowski space, it is easy: we just differentiate their components. But in a general space–time there is a problem because the coefficients in the transformation rules for vector and tensor components are generally not constant. A tensor that has constant components in one coordinate system will have varying components in another.

5.1 The Derivative of a Tensor

The derivatives of the components of a tensor do not themselves transform as tensor components. This is illustrated by the following examples.

Example 5.1

Let X^a be a vector field. Then

$$
\begin{aligned}
\tilde{\partial}_b \tilde{X}^a &= \frac{\partial x^c}{\partial \tilde{x}^b} \frac{\partial}{\partial x^c} \left(\frac{\partial \tilde{x}^a}{\partial x^d} X^d \right) \\
&= \frac{\partial x^c}{\partial \tilde{x}^b} \frac{\partial \tilde{x}^a}{\partial x^d} \partial_c X^d + \frac{\partial x^c}{\partial \tilde{x}^b} \frac{\partial^2 \tilde{x}^a}{\partial x^c \partial x^d} X^d ,
\end{aligned}
$$

where $\partial_b = \partial / \partial x^b$ and $\tilde{\partial}_b = \partial / \partial \tilde{x}^b$. The first term is the one required for a tensor transformation law; the second is the problem. In special relativity, where the coordinate transformations are all affine linear, it vanishes automatically. The difficulty in the general theory is that we now allow general, nonlinear coordinate transformations, for which it does not vanish.

Example 5.2

Let $x^a = x^a(\tau)$ be the worldline of a particle parametrized by proper time. Then

$$
\frac{\mathrm{d} x^a}{\mathrm{d}\tau} = \frac{\partial x^a}{\partial \tilde{x}^b} \frac{\mathrm{d} \tilde{x}^b}{\mathrm{d}\tau}
$$

which implies that the four-velocity components transform in the right way. But

$$
\frac{\mathrm{d}^2 x^a}{\mathrm{d}\tau^2} = \frac{\partial x^a}{\partial \tilde{x}^b} \frac{\mathrm{d}^2 \tilde{x}^b}{\mathrm{d}\tau^2} + \frac{\partial^2 x^a}{\partial \tilde{x}^b \partial \tilde{x}^c} \frac{\mathrm{d} \tilde{x}^b}{\mathrm{d}\tau} \frac{\mathrm{d} \tilde{x}^c}{\mathrm{d}\tau} .
$$

Again the second term is the obstruction to a nice transformation law. The obvious definition of four-acceleration does not give a vector.

The way out, which does lead to a tensor transformation law in both these cases, is to include an extra term involving the Christoffel symbols in the definition of the derivative. Under change of coordinates, the Christoffel symbols

$$
\Gamma^a_{bc} = \tfrac{1}{2} g^{ad} (\partial_b g_{dc} + \partial_c g_{ba} - \partial_a g_{bc})
$$

obey the transformation law

$$
\begin{aligned}
\Gamma^a_{bc} &= \frac{\partial x^a}{\partial \tilde{x}^d} \frac{\partial \tilde{x}^e}{\partial x^b} \frac{\partial \tilde{x}^f}{\partial x^c} \tilde{\Gamma}^d_{ef} + \frac{\partial x^a}{\partial \tilde{x}^d} \frac{\partial^2 \tilde{x}^d}{\partial x^b \partial x^c} \\
&= \frac{\partial x^a}{\partial \tilde{x}^d} \frac{\partial \tilde{x}^e}{\partial x^b} \frac{\partial \tilde{x}^f}{\partial x^c} \tilde{\Gamma}^d_{ef} - \frac{\partial^2 x^a}{\partial \tilde{x}^e \partial \tilde{x}^f} \frac{\partial \tilde{x}^e}{\partial x^b} \frac{\partial \tilde{x}^f}{\partial x^c} .
\end{aligned}
$$

The second term in the last line is exactly what we want to cancel the unwanted term in the first example. We define the *covariant derivative* of a vector field X^a by

$$
\nabla_b X^a = \partial_b X^a + \Gamma^a_{bc} X^c .
$$

We then have the following transformation law.

Proposition 5.3

The covariant derivative of a vector field transforms as a tensor of type $(1,1)$.

Proof

We express the covariant derivative in terms of new coordinates $\tilde{x}^a$:

$$
\begin{aligned}
& \partial_a X^b + \Gamma^b_{ad} X^d \\
&= \frac{\partial \tilde{x}^c}{\partial x^a} \frac{\partial}{\partial \tilde{x}^c} \left(\frac{\partial x^b}{\partial \tilde{x}^f} \tilde{X}^f \right) + \frac{\partial x^b}{\partial \tilde{x}^e} \frac{\partial \tilde{x}^f}{\partial x^a} \frac{\partial \tilde{x}^h}{\partial x^d} \tilde{\Gamma}^e_{fh} X^d - \frac{\partial^2 x^b}{\partial \tilde{x}^e \partial \tilde{x}^f} \frac{\partial \tilde{x}^e}{\partial x^a} \frac{\partial \tilde{x}^f}{\partial x^d} X^d \\
&= \frac{\partial \tilde{x}^c}{\partial x^a} \frac{\partial x^b}{\partial \tilde{x}^d} \left(\tilde{\partial}_c \tilde{X}^d + \tilde{\Gamma}^d_{ce} \tilde{X}^e \right) .
\end{aligned}
$$

$\square$

In a coordinate system such that $\partial_a g_{bc} = 0$ at the event $x^a = 0$, we have $\Gamma^a_{bc} = 0$ at $x^a = 0$ and hence that $\nabla_a X^b = \partial_a X^b$, although in general this holds only at the origin. We could have used this property to define the covariant derivative. That is, we could equally well define the covariant derivative by requiring that the value of $\nabla_a X^b$ at A should be the tensor that coincides with $\partial_a X^b$ in local inertial coordinates at A. Then the tensor transformation law would enable us to write down its components in a general coordinate system. It is a useful technique to define a tensor by giving its components in a particular coordinate system and then to use the transformation law backwards.

5.2 Parallel Transport

In taking the derivative a vector, we are comparing its values at nearby events, and finding the change. The coordinate derivatives of the components do not on their own give a good definition because the comparison is then simply of the components of the vector. The coefficients in the vector transformation law are not constant, so it is possible for a vector to have the same components at two different events in one coordinate system, but not in another. In one coordinate system, it appears to change between the events; in another it does not. By contrast, when we take the covariant derivative of X^a, we implicitly use *parallel transport* to compare the values of X at different events.

Let A and B be two nearby events with coordinates x^b and $x^b + \delta x^b$. To the first order in δx^b,

$$
\delta x^b \nabla_b X^a = \delta x^b \partial_b X^a + \delta x^b \Gamma^a_{bc} X^c
$$

$$= X^a(x + \delta x) - \left(X^a(x) - \delta x^b \Gamma^a_{bc} X^c(x) \right).$$

Thus the covariant derivative compares $X^a(x + \delta x)$, the value at B, with $X^a(x) - \delta x^b \Gamma^a_{bc} X^c$, which we think of as the result of displacing X^a from A to the 'most nearly parallel vector at B'.

Definition 5.4

The vector at B with components $X^a(A) - \delta x^b \Gamma^a_{bc} X^c(A)$ is said to be obtained by *parallel transport* of X^a from A to B.

In local inertial coordinates at A, we have $\Gamma = 0$ at A and the vector at B is the one with the same components as at A, to the first order in δx. It makes

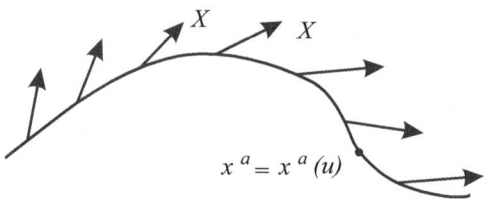

Figure 5.1 Parallel transport of X along the curve $x^a = x^a(u)$

more sense to express these ideas in terms of parallel transport along a curve: we then don't have to worry about infinitesimals.

Definition 5.5

A vector X is *parallel transported* or *parallel propagated* along a curve $x^a = x^a(u)$ whenever

$$\frac{\mathrm{d}X^a}{\mathrm{d}u} + \Gamma^a_{bc} \frac{\mathrm{d}x^b}{\mathrm{d}u} X^c = 0.$$

This is a set of ordinary differential equations for the components X^a as functions of the parameter u. It determines the X^as in terms of their values at the initial point of the curve.

Example 5.6

We can read the geodesic equation,

$$\ddot{x}^a + \Gamma^a_{bc}\dot{x}^b\dot{x}^c = 0,$$

as the statement that the four-velocity $\dot{x}^a$ is parallel propagated along the geodesic. This is the sense in which geodesics are curves in curved space–time which are 'as straight as possible'.

Parallel propagation around a closed curve need not return the vector to its starting value. This is a manifestation of curvature.

5.3 Covariant Derivatives of Tensors

The definition of the covariant derivative ∇_a extends to covectors by putting

$$\nabla_a \alpha_b = \partial_a \alpha_b - \alpha_c \Gamma^c_{ab}.$$

By a similar argument to that used in the case of vectors, this transforms as a tensor of type $(0, 2)$.

Exercise 5.1

Show that $\partial_a(\alpha_b Y^b) = (\nabla_a \alpha_b)Y^b + \alpha_b \nabla_a Y^b$. Note that $\alpha_b Y^b$ is a scalar, so the gradient covector on the left-hand side is well defined.

For a general tensor field, we define the covariant derivative by adding one gamma term for each upper index and subtracting one for each lower index. For example,

$$\nabla_a T^{bc}{}_d = \partial_a T^{bc}{}_d + \Gamma^b_{ae} T^{ec}{}_d + \Gamma^c_{ae} T^{be}{}_d - \Gamma^e_{ad} T^{bc}{}_e.$$

The first lower index on Γ in each term is a, the index on ∇. The rule for an upper index is: add a term $T\Gamma$, move the index to the upper position on Γ, and replace it by a dummy index, repeated as the second lower index on Γ. For a lower index, *subtract* a term $T\Gamma$, move the index to the second lower position on Γ, and replace it by a dummy index, repeated in the upper position on Γ. When there are no free indices, the covariant derivative is simply the partial derivative. Thus for a scalar f we write $\nabla_a f$ for $\partial_a f$. The covariant derivative of a tensor of type (p, q) is a tensor of type $(p, q + 1)$. The operation has the following properties.

(cd1) $\nabla_a(T^{\cdots}{}_{\cdots} + S^{\cdots}{}_{\cdots}) = \nabla_a T^{\cdots}{}_{\cdots} + \nabla_a S^{\cdots}{}_{\cdots}.$

(cd2) $\nabla_a(fT^{\cdots}{}_{\cdots}) = f\nabla_a T^{\cdots}{}_{\cdots} + (\nabla_a f)T^{\cdots}{}_{\cdots}\,.$

(cd3) $\nabla_a(T^{\cdots}{}_{\cdots}S^{\cdots}{}_{\cdots}) = \nabla_a(T^{\cdots}{}_{\cdots})S^{\cdots}{}_{\cdots} + T^{\cdots}{}_{\cdots}\nabla_a(S^{\cdots}{}_{\cdots})\,.$

(cd4) $\nabla_a T^{bc}{}_b$ is the same whether the contraction is done before or after the differentiation.

(cd5) The covariant derivative of the Kronecker delta vanishes because

$$\nabla_a \delta^b_c = \partial_a \delta^b_c + \Gamma^b_{ad}\delta^d_c - \Gamma^d_{ac}\delta^b_d = 0\,.$$

(cd6) For a scalar f, but not for a general tensor,

$$\nabla_a \nabla_b f = \partial_a \partial_b f - \partial_c f \Gamma^c_{ab} = \nabla_b \nabla_a f\,.$$

(cd7) The covariant derivative of the metric tensor vanishes, because

$$\begin{aligned}
\nabla_a g_{bc} &= \partial_a g_{bc} - g_{dc}\Gamma^d_{ab} - g_{bd}\Gamma^d_{ac} \\
&= \partial_a g_{bc} - \tfrac{1}{2}\{\partial_a g_{cb} + \partial_b g_{ac} - \partial_c g_{ab}\} \\
&\quad - \tfrac{1}{2}\{\partial_a g_{bc} + \partial_c g_{ab} - \partial_b g_{ac}\} \\
&= 0\,.
\end{aligned}$$

(cd8) $\nabla_a g^{bc} = 0$. This follows from (cd7) and

$$0 = \nabla_a(\delta^b_d) = \nabla_a(g^{bc}g_{cd}) = \nabla_a(g^{bc})g_{cd} + g^{bc}\nabla_a g_{cd}\,.$$

It follows from (cd7) and (cd8) that raising and lowering can be interchanged with covariant differentiation. For example, if X^a is a vector field, then $\nabla_a X_b$ is well defined. It does not matter whether you lower the index on the X before or after the differentiation.

Example 5.7 (Maxwell's equations)

In a curved space–time and in the absence of sources, these are

$$\nabla_a F^{ab} = 0, \qquad \nabla_a F_{bc} + \nabla_b F_{ca} + \nabla_c F_{ab} = 0$$

because these equations are covariant and reduce to the special relativity form in local inertial coordinates at a point. Gravity affects light through the Γs.

There is an important point here. It is not just that the equations coincide with Maxwell's equations in Minkowski space when there is no gravity; there are many other generalizations of the flat space–time equations with this property. It is that the equations in curved space–time are determined by the stronger requirement that they should involve only first derivatives and that they should reduce to the special relativity form in local inertial coordinates in the presence of gravity. If all that were required were that they should take correct form in Minkowski space, then it would be possible to add in other terms that vanished in the absence of gravity.

5.4 The Wave Equation

Suppose that u is a function on space–time. Then the partial derivatives $\partial_a u$ are the components of a covector, the gradient covector. We can define a vector field with components $\nabla^a u$ by putting

$$\nabla^a u = g^{ab} \partial_b u \, ;$$

that is, by raising the index. This is the *gradient vector*. The *wave operator* or *d'Alembertian* sends u to

$$\Box u = \nabla_a(\nabla^a u) = \partial_a\big(g^{ab}\partial_b u\big) + \Gamma^a_{ab} g^{bc}\partial_c u \, .$$

Now if A is a square matrix depending on the coordinates x^a, then

$$\partial_a \log \det A = \operatorname{tr}\big(A^{-1}\partial_a A\big)$$

(see the exercises at the end of this chapter). It follows that

$$\Gamma^b_{ab} = \tfrac{1}{2} g^{bd}\big(\partial_b g_{ad} + \partial_a g_{bd} - \partial_d g_{ab}\big) = \tfrac{1}{2} g^{bd}\partial_a g_{bd} = \partial_a \log \sqrt{|g|} \, ,$$

where g is the determinant of the matrix (g_{ab}). Hence

$$\Gamma^b_{ab} = \tfrac{1}{2}\partial_a \log |g| \, .$$

Therefore

$$\nabla_a \nabla^a u = \frac{1}{\sqrt{|g|}} \frac{\partial}{\partial x^a}\left(\sqrt{|g|}\, g^{ab}\frac{\partial u}{\partial x^b}\right) . \tag{5.1}$$

The operator on the right is invariant; it is independent of the choice of coordinates.

5.5 Connections

All that is needed to define the covariant derivative in a coordinate-independent way is that the Christoffel symbols should obey the transformation rule

$$\Gamma^a_{bc} = \frac{\partial x^a}{\partial \tilde{x}^d} \tilde{\Gamma}^d_{ef} \frac{\partial \tilde{x}^e}{\partial x^b} \frac{\partial \tilde{x}^f}{\partial x^c} + \frac{\partial x^a}{\partial \tilde{x}^d} \frac{\partial^2 \tilde{x}^d}{\partial x^b \partial x^c}\,.$$

A field of Γ^a_{bc}s with this transformation property is called a set of *connection coefficients*. The corresponding operator ∇ is called a connection: through parallel transport, it connects the spaces of vectors and tensors at nearby events.

Properties (cd1)–(cd5) are common to all connections; (cd6) holds only if the connection is torsion-free; that is, $\Gamma^c_{ab} = \Gamma^c_{ba}$; (cd7) holds in addition only for

$$\Gamma^a_{bc} = \tfrac{1}{2} g^{ad}(\partial_b g_{cd} + \partial_c g_{bd} - \partial_d g_{bc})\,. \tag{5.2}$$

This is the unique torsion-free connection for which the covariant derivative of the metric tensor vanishes. It is called the *Levi-Civita connection*.

It is easy to construct other examples of connections. If Γ^a_{bc} is one set of connection coefficients, for example, those of the Levi-Civita connection, and Q^a_{bc} is a tensor, then $\Gamma^a_{bc} + Q^a_{bc}$ is also a set of connection coefficients. All connections can be obtained in this way once one is given. From now on ∇ always denotes the Levi-Civita connection, defined by (5.2).

5.6 Curvature

In Minkowski space, there are global coordinate systems in which g_{ab} is constant. In such coordinates $\nabla_a = \partial_a$ and therefore $\nabla_a \nabla_b = \nabla_b \nabla_a$, when acting on vectors or tensors. So if in a general space–time, $\nabla_a \nabla_b \neq \nabla_b \nabla_a$ when acting on vectors, then we know that the metric cannot be reduced to the special relativity form by a coordinate change.

Proposition 5.8

For any metric g_{ab}, there is a tensor field $R_{abc}{}^d$ of type $(1,3)$ such that

$$\nabla_a \nabla_b X^d - \nabla_b \nabla_a X^d = R_{abc}{}^d X^c$$

for any four-vector field X.

The tensor $R_{abc}{}^d$ is called the *Riemann tensor* or *curvature tensor*.

Proof

From the definition of the Levi-Civita connection,

$$
\begin{aligned}
\nabla_a \nabla_b X^d &= \nabla_a(\partial_b X^d + \Gamma^d_{bc} X^c) \\
&= \partial_a \partial_b X^d + (\partial_a \Gamma^d_{bc}) X^c + \Gamma^d_{bc} \partial_a X^c \\
&\quad + \Gamma^d_{ae}(\partial_b X^e + \Gamma^e_{bc} X^c) - \Gamma^e_{ab}(\partial_e X^d + \Gamma^d_{ec} X^c).
\end{aligned}
$$

Hence

$$
(\nabla_a \nabla_b - \nabla_b \nabla_a) X^d = (\partial_a \Gamma^d_{bc} - \partial_b \Gamma^d_{ac} - \Gamma^d_{be} \Gamma^e_{ac} + \Gamma^d_{ae} \Gamma^e_{bc}) X^c,
$$

because the terms involving partial derivatives of X cancel. We define the expression in brackets to be $R_{abc}{}^d$. We must show that it is a tensor. The direct method is horrible. We know, however, that the left-hand side is a tensor. Hence, if we change coordinates,

$$
\begin{aligned}
\tilde{\nabla}_a \tilde{\nabla}_b \tilde{X}^d - \tilde{\nabla}_b \tilde{\nabla}_a \tilde{X}^d &= \frac{\partial x^p}{\partial \tilde{x}^a} \frac{\partial x^q}{\partial \tilde{x}^b} \frac{\partial \tilde{x}^d}{\partial x^s} R_{pqc}{}^s X^c \\
&= \frac{\partial x^p}{\partial \tilde{x}^a} \frac{\partial x^q}{\partial \tilde{x}^b} \frac{\partial x^r}{\partial \tilde{x}^c} \frac{\partial \tilde{x}^d}{\partial x^s} R_{pqr}{}^s \tilde{X}^c.
\end{aligned} \tag{5.3}
$$

Had we worked from the beginning in the new coordinates, we would have obtained

$$
(\tilde{\nabla}_a \tilde{\nabla}_b - \tilde{\nabla}_b \tilde{\nabla}_a) \tilde{X}^d = \tilde{R}_{abc}{}^d \tilde{X}^c, \tag{5.4}
$$

where $\tilde{R}_{abc}{}^d$ is defined in the same way as $R_{abc}{}^d$, but in the new coordinates. Because (5.3) and (5.4) hold for any X, we deduce that

$$
\tilde{R}_{abc}{}^d = \frac{\partial x^p}{\partial \tilde{x}^a} \frac{\partial x^q}{\partial \tilde{x}^b} \frac{\partial x^r}{\partial \tilde{x}^c} \frac{\partial \tilde{x}^d}{\partial x^s} R_{pqr}{}^s,
$$

which is the tensor transformation law. $\qquad \square$

Corollary 5.9

If there exists a vector field X such that $\nabla_a \nabla_b X^d \neq \nabla_b \nabla_a X^d$, then there does not exist a coordinate system in which the metric coefficients are constant.

5.7 Symmetries of the Riemann Tensor

The Riemann tensor encodes the second derivatives of the metric, and the first derivatives of the Christoffel symbols. Through the geodesic equation, the Γs give the 'acceleration due to gravity'. Thus the components $R_{abc}{}^d$ measure the difference in the acceleration between nearby points, which we identified as the 'real', frame-independent, effect of gravity.

A general four-index tensor has $4^4 = 256$ independent components. The Riemann tensor, however, has symmetries that reduce the number to 20. These are apparent from the form of the tensor in local inertial coordinates.

In terms of the connection coefficients

$$R_{abc}{}^d = \partial_a \Gamma^d_{bc} - \partial_b \Gamma^d_{ac} - \Gamma^d_{be}\Gamma^e_{ac} + \Gamma^d_{ae}\Gamma^e_{bc}.$$

Pick an event A and choose coordinates such that $\partial_a g_{bc} = 0$ at A. Then we also have $\Gamma^a_{bc} = 0$ and $\partial_a g^{bc} = 0$ at A. So, at the event A, but not elsewhere in general,

$$
\begin{aligned}
R_{abcd} &= g_{de}\partial_a(\Gamma^e_{bc}) - g_{de}\partial_b(\Gamma^e_{ac}) \\
&= \tfrac{1}{2}\partial_a(\partial_c g_{bd} + \partial_b g_{dc} - \partial_d g_{bc}) - \tfrac{1}{2}\partial_b(\partial_c g_{da} + \partial_a g_{dc} - \partial_d g_{ac}) \\
&= \tfrac{1}{2}(\partial_a\partial_c g_{bd} + \partial_b\partial_d g_{ac} - \partial_a\partial_d g_{bc} - \partial_b\partial_c g_{ad}).
\end{aligned}
\tag{5.5}
$$

From this we deduce that the Riemann tensor has the following symmetries.

(S1) $R_{abcd} = -R_{bacd}$

(S2) $R_{abcd} = R_{cdab}$

(S3) $R_{abcd} = -R_{abdc}$

(S4) $R_{abcd} + R_{bcad} + R_{cabd} = 0$.

The last of these can be expressed more simply by introducing special notation for dealing with calculations involving permutations of tensor indices.

Bracket notation

For a general covariant tensor with p lower indices, we define

$$T_{[ab...c]} = \frac{1}{p!} \sum_{\text{perms}} \text{sign}\,(\sigma) T_{\sigma(a)\sigma(b)...\sigma(c)}$$

$$T_{(ab...c)} = \frac{1}{p!} \sum_{\text{perms}} T_{\sigma(a)\sigma(b)...\sigma(c)},$$

where the sums are over the permutations σ of p objects, and $\text{sign}\,(\sigma)$ is 1 or -1 as σ is even or odd. For example,

$$
\begin{aligned}
T_{[ab]} &= \tfrac{1}{2}(T_{ab} - T_{ba}) \\
T_{(ab)} &= \tfrac{1}{2}(T_{ab} + T_{ba}) \\
T_{[abc]} &= \tfrac{1}{6}(T_{abc} + T_{bca} + T_{cab} - T_{bac} - T_{acb} - T_{cba}) \\
T_{(abc)} &= \tfrac{1}{6}(T_{abc} + T_{bca} + T_{cab} + T_{bac} + T_{acb} + T_{cba}).
\end{aligned}
$$

The same definitions apply to brackets on a subset of the indices and to brackets on upper indices. For example

$$T^{[ab](cd)} = \tfrac{1}{4}\left(T^{abcd} - T^{bacd} + T^{abdc} - T^{badc}\right).$$

There is a possibility of ambiguity over the order of the operations if two sets of brackets partially overlap, as, for example, in the expression $T_{[a(bc]d)}$. So partial overlaps are forbidden. Nested brackets, however, are unambiguous, although they can always be simplified because

$$T_{[\dots(\dots)\dots]} = 0 = T_{(\dots[\dots]\dots)}, \quad T_{[\dots[\dots]\dots]} = T_{[\dots\dots\dots]}, \quad T_{(\dots(\dots)\dots)} = T_{(\dots\dots\dots)}.$$

Example 5.10

The symmetries of the contravariant metric g^{ab} and of the alternating tensor ε_{abcd} can be expressed, respectively, as

$$g^{[ab]} = 0, \qquad \varepsilon_{abcd} = \varepsilon_{[abcd]}.$$

Maxwell's equations without sources are

$$\nabla_a F^{ab} = 0, \qquad \nabla_{[a} F_{bc]} = 0.$$

The second is an automatic consequence of the relationship $F_{ab} = 2\nabla_{[a}\Phi_{b]}$ between the electromagnetic field F_{ab} and the four-potential Φ_a. In fact, it is locally equivalent to the existence of the four-potential.

With this notation, the fourth symmetry (S4) of the Riemann tensor reads

$$R_{[abc]d} = 0.$$

The Riemann tensor also automatically satisfies a differential identity—the *Bianchi identity*—as a consequence of the fact that it is derived from the a metric and its derivatives. It is analogous to the vanishing of $\nabla_{[a} F_{bc]}$ as a consequence of the existence of the four-potential.

Proposition 5.11 (The Bianchi identity)

$$\nabla_{[a} R_{bc]d}{}^{e} = 0.$$

Proof

Choose coordinates such that $\Gamma^a_{bc} = 0$ at an event. We have

$$\nabla_a R_{bcd}{}^e = \partial_a \partial_b \Gamma^e_{cd} - \partial_a \partial_c \Gamma^e_{bd} + \text{terms in } \Gamma \partial \Gamma \text{ and } \Gamma \Gamma \Gamma .$$

Because the first term on the right-hand side is symmetric in ab and the second in ac, and because the other terms vanish at the event, we have

$$\nabla_{[a} R_{bc]d}{}^e = 0$$

at the event in this coordinate system. However, this is a tensor equation, so it is valid in every coordinate system. □

The Riemann tensor encodes the observable, frame-independent aspects of the gravitational field. In the next two sections, we consider two interpretations of the tensor that allow us to relate its components directly to physical observations.

5.8 Geodesic Deviation

The first interpretation is in terms of relative acceleration of nearby particles in free-fall. Consider an observer O with worldline ω. Let τ denote the proper time along ω and let

$$V^a = \frac{\mathrm{d}x^a}{\mathrm{d}\tau}$$

denote the four-velocity of O. We want to find the acceleration of a nearby particle in free-fall in terms of its four-velocity and position relative to O. To do this we need a tool, a derivative operator that measures the rate of change of vectors and tensors along ω.

The operator D

Let $Y^a(\tau)$ be a vector field. Its covariant derivative DY^b along ω is defined by the following equivalent expressions,

$$\begin{aligned}
DY^b &= V^a \nabla_a Y^b \\
&= \frac{\mathrm{d}x^a}{\mathrm{d}\tau} \partial_a Y^b + \Gamma^b_{ac} V^a Y^c \\
&= \frac{\mathrm{d}Y^b}{\mathrm{d}\tau} + \Gamma^b_{ac} V^a Y^c .
\end{aligned}$$

The first makes it clear that DY^a is a well-defined vector at each point of ω. The last, that the values of DY^a along ω depend only on the values of $Y^a(\tau)$ along ω, so DY^a makes sense for vector fields that are defined only along ω. Note that $DY^a = 0$ is the equation of parallel transport.

The operator extends in a natural way to tensor fields. For example,

$$DT^a_b = \frac{\mathrm{d}T^a_b}{\mathrm{d}\tau} + \Gamma^a_{cd}V^cT^d_b - \Gamma^d_{cb}V^cT^a_d.$$

The definition makes sense for any timelike worldline. But if the observer is in free-fall, so ω is a geodesic, then $D = \mathrm{d}/\mathrm{d}\tau$ at the origin of local inertial coordinates in which the observer is instantaneously at rest.

Now imagine a cloud of particles in free-fall. Let us suppose that an observer O is travelling with one of the particles, and that this particle has worldline ω. Suppose that the observer looks at a nearby particle and measures its position in local inertial coordinates. In special relativity, it will move in a straight line at constant speed, and will have no acceleration. What happens in a gravitational field?

The four-velocities of the particles form a vector field V^a. Because the individual particle worldlines are geodesic,

$$V^b\nabla_bV^a = DV^a = \frac{\mathrm{d}V^a}{\mathrm{d}\tau} + \Gamma^a_{bc}V^bV^c = 0\,.$$

Pick out a particle P near O, and at each event on ω, let Y^a be the four-vector joining the event to a simultaneous event at P. Because P is 'near' O, Y is small. We ignore second-order terms in its components.

In the local inertial coordinates in which O is instantaneously at rest, Y has components $(0, \boldsymbol{y})$, where $\boldsymbol{y}$ is the position of P. If ω is given by $x^a = x^a(\tau)$ in general coordinates, then P's worldline is

$$x^a(\tau) + Y^a(\tau) + O(2)\,, \tag{5.6}$$

where $O(2)$ denotes second-order and smaller terms in the coordinates of P, and τ is the proper time along the worldline of O. Now the proper time separation $\mathrm{d}\tau$ between two nearby events $x^a(\tau)$ and $x^a(\tau + \delta\tau)$ on the worldline of O is the same to the second order in $\boldsymbol{y}$ as the proper time between the corresponding events on the worldline of P with coordinates

$$x^a(\tau) + Y^a(\tau) \qquad \text{and} \qquad x^a(\tau + \delta\tau) + Y^a(\tau + \delta\tau)\,.$$

Within our approximation, therefore, τ is also the proper time along P's worldline.

We note that Y^a is a vector field along ω and that it is orthogonal to V^a in the sense that $V^aY_a = 0$, because $Y = (0, \boldsymbol{y})$ and $V = (1, 0)$ in the local inertial coordinates in which O is instantaneously at rest. Because $DV^a = 0$, we also have

$$0 = D(V_a Y^a) = V_a DY^a \qquad \text{and} \qquad 0 = D(V_a DY^a) = V_a D^2 Y^a\,.$$

In the local rest frame of O at an event on ω, the four-velocity of O is $(1, \mathbf{0})$, and the vectors Y^a, DY^a, and $D^2 Y^a$ are, respectively, $(0, \boldsymbol{y})$, $(0, \boldsymbol{u})$, and $(0, \boldsymbol{a})$, where $\boldsymbol{u}$ is the relative velocity of P to O and $\boldsymbol{a}$ is the relative acceleration.

We are interested in the relative acceleration, and therefore in $D^2 Y^a$. We want to express this in terms of the curvature. The key to this is the following result.

Proposition 5.12

$DY^a = Y^b \nabla_b V^a$.

Proof

We know from (5.6) that

$$V^a(P) = \frac{\mathrm{d}x^a}{\mathrm{d}\tau} + \frac{\mathrm{d}Y^a}{\mathrm{d}\tau} + O(2) = V^a(O) + \frac{\mathrm{d}Y^a}{\mathrm{d}\tau} + O(2)\,.$$

On the other hand, by expanding to the first order in the separation of O and P,

$$V^a(P) = V^a(O) + Y^c \partial_c V^a + O(2)\,.$$

Therefore $\mathrm{d}Y^a/\mathrm{d}\tau = Y^c \partial_c V^a$. It follows that

$$DY^a = \frac{\mathrm{d}Y^a}{\mathrm{d}\tau} + \Gamma^a_{bc} V^b V^c = Y^c \partial_c V^a + \Gamma^a_{bc} V^b Y^c = Y^b \nabla_b V^a\,,$$

which is the result we need.   □

Now we can derive the *equation of geodesic deviation* or *Jacobi equation*, which is central to the physical interpretation of curvature.

$$\begin{aligned}
D^2 Y^d &= D(Y^b \nabla_b V^d) \\
&= (DY^b) \nabla_b V^d + Y^b D(\nabla_b V^d) \\
&= (Y^a \nabla_a V^b) \nabla_b V^d + Y^b V^a \nabla_a \nabla_b V^d \\
&= Y^a (\nabla_a V^b) \nabla_b V^d + Y^b V^a \nabla_b \nabla_a V^d + R_{abc}{}^d V^a Y^b V^c\,. \quad (5.7)
\end{aligned}$$

But

$$V^a \nabla_b \nabla_a V^d = \nabla_b (V^a \nabla_a V^d) - (\nabla_b V^a)(\nabla_a V^d) = -(\nabla_b V^a)(\nabla_a V^d)$$

because $V^u \nabla_a V^d = 0$ by the geodesic equation. Therefore the first two terms in the last line of (5.7) cancel, and

$$D^2 Y^d = R_{abc}{}^d V^a Y^b V^c,$$

which is the geodesic deviation equation. It gives the relative acceleration of nearby particles in free-fall in terms of their separation and of the curvature tensor.

5.9 Geodesic Triangles*

The Gaussian curvature of a surface determines the excess of the sum of the angles of a geodesic triangle over π. There is an analogous interpretation of the Riemann tensor in space–time, which gives a direct way to understand its physical meaning. In this case, the geodesics are free-particle worldlines, and the angles are the *rapidities* of the particles.

Rapidity

Consider two particle worldlines through an event A. Suppose that the four-velocities of the particles at A are U and V. Then the *rapidity* θ of one particle relative to the other is defined by

$$\cosh \theta = U_a V^a.$$

If one particle is at rest in local inertial coordinates at A, and the other has speed v, then

$$\cosh \theta = \gamma(v) = \frac{1}{\sqrt{1 - v^2}}.$$

Rapidity is the space–time analogue of 'angle'. The relativistic addition formula for velocities translates into additivity of rapidities, in the following sense. Suppose that A is on the worldlines of three particles O, P, and Q. Let θ_{OP} denote the rapidity of O relative to P and so on. If the particles' respective four-velocities U, V, W at A are coplanar at A, with V a linear combination of U and W with positive coefficients, then

$$\theta_{OQ} = \theta_{OP} + \theta_{PQ} \qquad V \sinh \theta_{OQ} = W \sinh \theta_{OP} + U \sinh \theta_{PQ}. \tag{5.8}$$

See Figure (5.2). If the relative speeds are small, then (5.8) reduces in the limit to the classical velocity addition formula $v_{OQ} = v_{OP} + v_{PQ}$, where the vs denote relative speed.

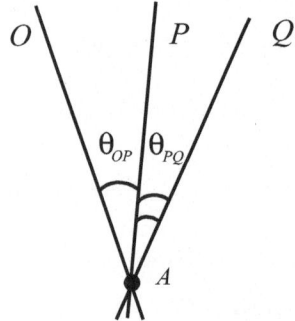

Figure 5.2   The addition of rapidity

Exercise 5.2

Establish the second identity in (5.8).

We need a variant of (5.8) to derive our interpretation of the Riemann tensor. Suppose that Q' and Q'' are two further particles with respective four-velocities W' and W'' at A. Suppose further that $X' = W' - W$ and $X'' = W'' - W$ are small, so that we can ignore second-order terms in their components. Then

$$X_a'W^a = 0, \qquad X_a''W^a = 0\,,$$

and

$$(\theta_{OQ'} - \theta_{OQ})\sinh\theta_{OQ} = X_a'U^a$$

to within the approximation, by applying Taylor's theorem to the left-hand side of

$$\cosh\theta_{OQ'} = W_a'U^a = \cosh\theta_{OQ} + X_a'U^a\,.$$

We also have a similar formula relating $\theta_{PQ''}$, θ_{PQ}, and $X_a''V^a$. By appealing to the second identity in (5.8), we conclude that to within our approximation

$$
\begin{aligned}
\theta_{OQ'} - \theta_{OP} - \theta_{PQ''} &= \frac{X_a'U^a}{\sinh\theta_{OQ}} - \frac{X_a''V^a}{\sinh\theta_{PQ}}\\
&= \frac{U^a(X_a' - X_a'')}{\sinh\theta_{OQ}}\\
&= \frac{U^a(W_a' - W_a'')}{\sinh\theta_{OQ}}\,.
\end{aligned}
\tag{5.9}
$$

We now consider the following situation, mirroring a geodesic triangle on a surface; see Figure 5.3. Suppose that O and P are free particles whose worldlines

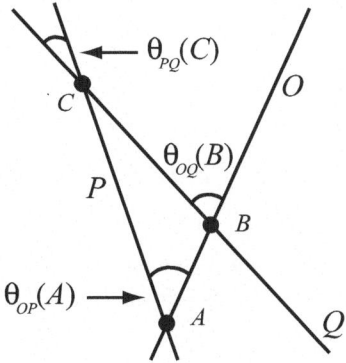

Figure 5.3 A geodesic triangle

pass through A. Let B be an event at proper time λ after A on the worldline of O and let C be an event at proper time μ after A on the worldline of P. Suppose that Q is a third particle whose worldline passes through B and C, with B to the past of C. Let $\theta_{OP}(A)$, $\theta_{OQ}(B)$, and $\theta_{PQ}(C)$ denote, respectively, the rapidity of O relative to P at A, and so on.

In Minkowski space, we have

$$\theta_{OQ}(B) - \theta_{OP}(A) - \theta_{PQ}(C) = 0\,,$$

by translating the worldline of Q, which is a straight line, to a parallel line through A, and by appealing to (5.8). The formula also holds in Euclidean geometry if we interpret $\theta_{OP}(A)$ as the interior angle of a triangle ABC at A, $\theta_{OQ}(B)$ as the exterior angle at B, and $\theta_{PQ}(C)$ as the interior angle at C. It is simply the statement that the sum of the interior angles is π. On a surface, the left-hand side is equal to the integral of the Gaussian curvature over the triangle. In curved space–time, we have the following.

Proposition 5.13

To the second order in λ, μ,

$$\theta_{OQ}(B) - \theta_{OP}(A) - \theta_{PQ}(C) = -\frac{\lambda\mu R_{abcd}U^a V^b U^c V^d}{2\sinh\theta_{OP}}\,,$$

where U and V are the four-velocities of O and P and the right-hand side is evaluated at A.

Proof

Let $x^a = x^a(\tau)$ denote the four-velocity of the particle Q, parametrized by proper time τ. By differentiating the geodesic equation, and by applying Taylor's theorem,

$$\dot{x}^a(\tau) = \dot{x}^a - \tau \dot{x}^b \dot{x}^c \Gamma^a_{bc} - \tfrac{1}{2}\tau^2 \dot{x}^b \dot{x}^c \dot{x}^d \partial_d \Gamma^a_{bd} + \tau^2 \dot{x}^c \dot{x}^d \dot{x}^e \Gamma^a_{bc} \Gamma^b_{de} + O(\tau^3)$$

as $\tau \to 0$, where $\dot{x}^a = \dot{x}^a(0)$.

Now choose the coordinates to be local inertial coordinates at A and let ν denote the proper time along the geodesic from B to C. Let $W^a(B)$ and $W^a(C)$ denote the components of the four-velocity of Q at B and C, respectively. We have

$$\Gamma^a_{bc}(A) = 0, \qquad \Gamma^a_{bc}(B) = \lambda U^d \partial_d \Gamma^a_{bc} + O(\lambda^2),$$

where the derivative of the Christoffel symbol is evaluated at A. Hence by using the geodesic equation for Q, to the second order,

$$W^a(C) = W^a(B) - \lambda \nu U^d W^b W^c \partial_d \Gamma^a_{bc} - \tfrac{1}{2}\nu^2 W^d W^b W^c \partial_d \Gamma^a_{bc}.$$

In the second and third terms on the right, it does not matter at which events the four-velocity components are evaluated. We also have, to the first order,

$$\mu V^a = \lambda U^a + \nu W^a.$$

Now let $\tilde{W}'^a$ and $\tilde{W}''^a$ denote the components at A of the vectors obtained by parallel transport of $W^a(B)$ and $W^a(C)$ along the worldlines of O and P, respectively.

By differentiating the parallel transport equation

$$\frac{\mathrm{d}W^a}{\mathrm{d}\tau} + \Gamma^a_{bc} U^b W^c = 0,$$

with respect to proper time along the worldline of O, and by using the fact that $\Gamma^a_{bc} = 0$ at A, and

$$\frac{\mathrm{d}\Gamma^a_{bc}}{\mathrm{d}\tau} = U^d \partial_d \Gamma^a_{bc},$$

we have

$$W^a(B) = W'^a - \tfrac{1}{2}\lambda^2 U^d U^b W^c \partial_d \Gamma^a_{bc},$$

by Taylor's theorem, again up to the second order. With the similar formula for transport along the worldline of P, we deduce that, to the second order,

$$
\begin{aligned}
W''^a - W'^a \\
&= \tfrac{1}{2}\left(\mu^2 V^d V^b - \lambda^2 U^d U^b - 2\lambda \nu U^d W^b - \nu^2 W^d W^b\right) W^c \partial_d \Gamma^a_{bc} \\
&= \tfrac{1}{2}\lambda \mu (U^b V^d - U^d V^b) W^c \partial_d \Gamma^a_{bc} \\
&= \tfrac{1}{2}\lambda \mu R_{dbc}{}^a U^b V^d W^c,
\end{aligned}
$$

by using the formula (5.5) for the curvature in local inertial coordinates.

Because the inner product is preserved by parallel transport, the rapidity of Q relative to O at B is the same as the rapidity of a particle Q' with four-velocity W'^a with respect to O at A. Similarly the rapidity of Q relative to P at C is the same as the rapidity of a particle Q'' with four-velocity W''^a with respect to P at A. We conclude that

$$\theta_{OQ}(B) - \theta_{OP}(A) - \theta_{PQ}(C) = \theta_{OQ'} - \theta_{OP} - \theta_{PQ''}.$$

But by using (5.9),

$$\theta_{OQ'} - \theta_{OP} - \theta_{PQ''} = -\frac{\lambda\mu R_{dbca}U^bV^dW^cU^a}{2\sinh\theta_{OQ}} = -\frac{\lambda\mu R_{dbca}U^bV^dV^cU^a}{2\sinh\theta_{OP}}.$$

to within our approximation. The proposition follows. $\square$

This result gives us a direct physical interpretation of the curvature quantity $R_{abcd}U^aV^bU^cV^d$ for two four velocities. Imagine two observers O, P with four-velocities U^a and V^a at an event A. After proper time λ on measured on O's clock, O throws a ball Q to P, who catches it after proper time μ, measured on P's worldline. The ball is thrown at event B and caught at event C. The observers can measure (i) their relative speed A, (ii) the speed of the ball relative to the first observer at B, and (iii) the speed of the ball relative to the second observer at C. They can therefore between them compute the quantity $\theta_{OQ}(B) - \theta_{OP}(A) - \theta_{PQ}(C)$, and hence measure $R_{abcd}U^aV^bU^cV^d$.

EXERCISES

5.3. Let ∇ be any torsion-free connection. Show that if X, Y are vector fields, then $X^b\partial_bY^a - Y^b\partial_bX^a = X^b\nabla_bY^a - Y^b\nabla_bX^a$.

5.4. Show that if α is a covector field, then $\partial_a\alpha_b - \partial_b\alpha_a$ transforms as a tensor of type $(0,2)$. This tensor is called the *exterior derivative* of α and is denoted by $d\alpha$. Show that the components of $d\alpha$ are also given by $\nabla_a\alpha_b - \nabla_b\alpha_a$ for any torsion-free connection ∇.

5.5. Let ∇ be a torsion-free connection, given by

$$\nabla_aX^b = \partial_aX^b + \Gamma^b_{ac}X^c.$$

Show that if $\nabla_ag_{bc} = 0$, then

$$\partial_ag_{bc} = K_{bac} + K_{cab},$$

where $K_{bac} = g_{bd}\Gamma^d_{ac}$. Deduce that ∇ is the Levi-Civita connection.

5.6. Establish the transformation law

$$\Gamma^a_{bc} = \tilde{\Gamma}^d_{ef}\frac{\partial x^a}{\partial \tilde{x}^d}\frac{\partial \tilde{x}^e}{\partial x^b}\frac{\partial \tilde{x}^f}{\partial x^c} + \frac{\partial x^a}{\partial \tilde{x}^d}\frac{\partial^2 \tilde{x}^d}{\partial x^b \partial x^c}$$

for the Christoffel symbols

$$\Gamma^a_{bc} = \tfrac{1}{2}g^{ad}(\partial_b g_{cd} + \partial_c g_{bd} - \partial_d g_{bc})$$

by direct calculation.

5.7. Let A and B be 4×4 matrices. Show that to the first order in ϵ,

$$\det (I + \epsilon B) = 1 + \epsilon \operatorname{tr} B \,;$$

and that $\det (A + \epsilon B) = \det A \det (I + \epsilon A^{-1}B)$.

Let g denote the determinant of the matrix (g_{ab}) of metric coefficients. Show that

$$\partial_a \log |g| = g^{bc}\partial_a g_{bc}.$$

Deduce that the Christoffel symbols satisfy $\Gamma^b_{ab} = \partial_a \log |g|^{1/2}$.

Show that in general coordinates x^a on Minkowski space, the wave equation is $\nabla_a \nabla^a u = 0$, where ∇ is the Levi-Civita connection of the Minkowski space metric g_{ab}. Show that this can be written as

$$\partial_a(|g|^{1/2}g^{ab}\partial_b u) = 0.$$

Hence write down the wave equation in spherical polar coordinates.

5.8. Let X be a vector field and let ∇ be the Levi-Civita connection. Show that if there exists a coordinate system in which $(X^a) = (1,0,0,0)$ (everywhere) and $\partial_0 g_{bc} = 0$, then $\nabla_a X_b + \nabla_b X_a = 0$ in every coordinate system.

What is the corresponding result if $\partial_0 g_{bc} = 0$ is replaced by $\partial_0 g_{bc} = f g_{bc}$ for some scalar field f?

5.9. A tensor in space–time satisfies $T_{abcde} = T_{[abcde]}$. Show that T_{abcde} is zero.

5.10. A tensor T_{ab} is symmetric if $T_{ab} = T_{(ab)}$. In n-dimensional space, it has n^2 components, but only $\tfrac{1}{2}n(n+1)$ of these can be specified independently, for example, the components T_{ab} for $a \le b$. How many independent components do the following tensors have?

(a) F_{ab} with $F_{ab} = F_{[ab]}$.

(b) A tensor of type $(0,k)$ such that $T_{ab...c} = T_{[ab...c]}$. Distinguish the cases $k \le 4$ and $k > 4$.

(c) R_{abcd} with $R_{abcd} = R_{[ab]cd} = R_{ab[cd]}$.

(d) R_{abcd} with $R_{abcd} = R_{[ab]cd} = R_{ab[cd]} = R_{cdab}$.

5.11. Show that symmetries (S1), (S3), and (S4) of the Riemann tensor imply (S2).

5.12. Show that for any covector field X_a,

$$\nabla_a \nabla_b X_c - \nabla_b \nabla_a X_c = -R_{abc}{}^d X_d .$$

Here ∇ is the Levi-Civita connection in space–time. Show that if $\nabla_{(a} X_{b)} = 0$, then $\nabla_a \nabla_b X_c = R_{bca}{}^d X_d$. Deduce that X^a satisfies the equation of geodesic deviation along any geodesic.

5.13. Let A be a covector field. Define $F_{ab} = \nabla_a \Phi_b - \nabla_b \Phi_a$. Show that the second of Maxwell's equations ($\nabla_{[a} F_{bc]} = 0$) is satisfied for any Φ, but that the first ($\nabla_a F^{ab} = 0$) holds if and only if

$$\Box \Phi_a - \nabla_a (\nabla_b \Phi^b) = -R_{ab} \Phi^b ,$$

where $\Box = \nabla_a \nabla^a$.

5.14. Show that if f is a function such that $(\nabla_a f)(\nabla^a f)$ is constant, then $X^a = \nabla^a f$ satisfies $X^a \nabla_a X^b = 0$; that is, that the integral curves of X, which are the solutions of $dx^a / d\tau = X^a$, are geodesics.

5.15. Write down the geodesic equations for the metric

$$ds^2 = dudv + \log(x^2 + y^2)du^2 - dx^2 - dy^2$$

($0 < x^2 + y^2 < 1$). Show that $K = x\dot{y} - y\dot{x}$ is a constant of the motion.

By considering an equivalent problem in Newtonian mechanics, show that no geodesic on which $K \neq 0$ can reach $x^2 + y^2 = 0$.

5.16. Show that for any tensor field $T^{ab}{}_c$,

$$(\nabla_a \nabla_b - \nabla_b \nabla_a) T^{ef}{}_k = R_{abc}{}^e T^{cf}{}_k + R_{abc}{}^f T^{ec}{}_k - R_{abk}{}^c T^{ef}{}_c .$$

Einstein's Equation

The relative acceleration of two nearby particles in free-fall is determined by the equation of geodesic deviation

$$D^2 Y^d = R_{abc}{}^d V^a V^c Y^b \,.$$

From the viewpoint of an observer travelling with the first particle, the acceleration of the second is a linear function of its position. From this starting point, we are led to Einstein's equation as the successor to Poisson's equation in the classical theory of gravity.

6.1 Tidal Forces

In local inertial coordinates in which the observer is instantaneously at rest, $V = (1, \mathbf{0})$ and $Y = (0, \mathbf{y})$, where $\mathbf{y}$ is the position vector of the second particle. The acceleration is

$$\mathbf{a} = -M\mathbf{y} \,,$$

where M is the 3×3 symmetric matrix with entries

$$M_{ij} = R_{0i0j} \,,$$

the symmetry following from the symmetries of the Riemann tensor.

What is the corresponding result in Newtonian gravity? Consider a cloud of particles in free-fall. The acceleration of each particle is given by $\ddot{\mathbf{r}} = -\boldsymbol{\nabla}\phi$.

So by Taylor's theorem, the relative acceleration $\boldsymbol{a}$ of two nearby particles O and P has components

$$a_i = (-\partial_i \phi)_P - (-\partial_i \phi)_O = -y_j \partial_j \partial_i \phi + O(2) \,, \tag{6.1}$$

where $\boldsymbol{y}$ is the vector from O to P and $O(2)$ denotes second-order terms in the components of $\boldsymbol{y}$. The second derivatives are evaluated at O, and there is a sum over $j = 1, 2, 3$.

Tides

One context in which (6.1) has a familiar interpretation is in the theory of tides. If O and P are in the moon's gravitational field, with the line joining them directed towards the moon, then (6.1) gives a relative acceleration towards the moon. This is true whichever particle is in the lead, because interchanging the particles reverses the sign of $\boldsymbol{y}$ and therefore of $\boldsymbol{a}$. So if we think of O as at the centre of the earth, and of P as a mass of water on the surface, then the moon's gravity gives rise to a *tidal force* on P acting away from O. This is true whether P is on the surface directly under the moon or on the opposite side of the earth. The tidal force raises two humps in the ocean, one under the moon and one on the opposite side of the earth. As the earth rotates, the humps move round, giving two high tides each day. We can trace the reason that there are two high tides to the linearity of (6.1) in $\boldsymbol{y}$. Before Newton, even Galileo's explanation of the tidal cycle was confused, and erroneous.

When our observer in curved space–time looks at the acceleration of nearby particles, and interprets his observations in terms of Newtonian theory, he imagines that he is in a gravitational field with potential ϕ such that

$$M_{ij} = \partial_i \partial_j \phi = R_{0i0j} \,.$$

Now in empty space, Poisson's equation reduces to $\nabla^2 \phi = 0$. That is, $\partial_i \partial_i \phi = 0$ or, with the observer's Newtonian interpretation, $\operatorname{tr} M = 0$. Thus in general relativity, we should have $R_{0i0i} = 0$ in empty space. Because $R_{0000} = 0$ by the symmetries of the Riemann tensor, an equivalent statement is

$$R_{abc}{}^b V^a V^c = 0 \,.$$

As this must hold for every four-velocity V, we are led to *Einstein's vacuum equation*

$$R_{ab} = 0 \,, \tag{6.2}$$

where R_{ab} is the *Ricci tensor*, defined by

$$R_{ab} = R_{acb}{}^c \,.$$

The vacuum equation is in fact ten equations, one for each of the ten independent components of the symmetric tensor R_{ab}, in ten unknowns, the ten independent components of the metric g_{ab}. The equations are nonlinear, as anticipated. The use of the summation convention makes them look very simple. Written out explicitly without this notation, the expression for each component of R_{ab} would contain over a thousand terms. Not surprisingly, therefore, it is not easy to find solutions.

We note two justifications for the vacuum equation. It reduces to the Newtonian equation in the weak field limit, and it has a solution, the *Schwarzschild solution*, analogous to $\phi = -Gm/r$, which encodes the inverse square law of gravity in Newtonian theory.

6.2 The Weak Field Limit

The reduction to Newtonian theory occurs when the metric is close to that of Minkowski space, so the gravitational field is 'weak', and when the configuration is nearly static, so the metric is not varying rapidly with time. It begins with the assumption that

$$g_{ab} = m_{ab} + h_{ab} \,,$$

where $m_{ab} = \mathrm{diag}(1, -1, -1, -1)$ is the Minkowski space metric in an inertial coordinate system x^a, and h_{ab} is small and slowly varying. 'Small' means that we can ignore any terms that involve products of two or more components of h_{ab} or its derivatives; 'slowly varying' means that we can ignore terms involving derivatives of h_{ab} with respect to the time coordinate $t = x^0$.

To obtain the vacuum equation in this case, we first have to find the contravariant metric g^{ab}, defined by $g^{ab}g_{bc} = \delta^a_c$. In our approximation, it is given by

$$g^{ab} = m^{ab} - m^{ac}m^{bd}h_{cd} \,,$$

where $m^{ab} = \mathrm{diag}(1, -1, -1, -1)$ is the contravariant Minkowski space metric, as is verified by the following calculation, in which the product of two h-terms is ignored.

$$g^{ab}g_{bc} = (m^{ab} - m^{ad}m^{be}h_{de})(m_{bc} + h_{bc}) = m^{ab}m_{bc} - m^{ad}h_{dc} + m^{ab}h_{bc} = \delta^a_c \,.$$

There is an immediate possibility for confusion here because we are dealing simultaneously with two metrics, m and g, so for the moment we avoid raising and lowering indices with either.

The Christoffel symbols are given to within our approximation by

$$\Gamma^a_{bc} = \tfrac{1}{2}g^{ad}(\partial_b g_{cd} + \partial_c g_{bd} - \partial_d g_{bc}) = \tfrac{1}{2}m^{ad}(\partial_b h_{cd} + \partial_c h_{bd} - \partial_d h_{bc}) \qquad (6.3)$$

and therefore the approximate Riemann tensor is

$$
\begin{aligned}
R_{abc}{}^{d} &= \partial_a \Gamma^d_{bc} - \partial_b \Gamma^d_{ac} \\
&= \tfrac{1}{2} m^{de} (\partial_a \partial_c h_{be} + \partial_b \partial_e h_{ac} - \partial_a \partial_e h_{bc} - \partial_b \partial_c h_{ae}).
\end{aligned}
\tag{6.4}
$$

Note that the $\Gamma\Gamma$-terms in the definition of the Riemann tensor have been dropped because they involve products of derivatives of the components of h.

Consider the motion of a slow-moving particle with worldline $x^a = x^a(t)$. We have

$$
\frac{\mathrm{d}x^0}{\mathrm{d}t} = 1, \qquad \frac{\mathrm{d}x^1}{\mathrm{d}t} = u_1, \qquad \frac{\mathrm{d}x^2}{\mathrm{d}t} = u_2, \qquad \frac{\mathrm{d}x^3}{\mathrm{d}t} = u_3,
$$

where (u_1, u_2, u_3) is the velocity in the inertial coordinates on Minkowski space. The 'slow moving' assumption is that the velocity is also small, so we ignore products of the u_is with each other and with the h_{ab}s and their derivatives. In particular, we have $\gamma(u) \sim 1$ and so we can identify the coordinate time t with the proper time τ along the worldline, and so approximate the four-velocity by

$$
(V^a) = (1, u_1, u_2, u_3).
$$

The geodesic equation is

$$
\frac{\mathrm{d}^2 x^a}{\mathrm{d}\tau^2} + \Gamma^a_{bc} \frac{\mathrm{d}x^b}{\mathrm{d}\tau} \frac{\mathrm{d}x^c}{\mathrm{d}\tau} = 0.
$$

Because we ignore products of the spatial components of the four-velocity with the Christoffel symbols, this is approximated by

$$
\frac{\mathrm{d}^2 x^a}{\mathrm{d}\tau^2} + \Gamma^a_{00} = 0.
$$

We also ignore terms involving time (x^0) derivatives of the metric components. Therefore Γ^a_{00} is only significant for $a \neq 0$. We have

$$
\Gamma^1_{00} = -\tfrac{1}{2} m^{11} \partial_1 h_{00} = \tfrac{1}{2} \partial_1 h_{00}
$$

and so on. The first component of the geodesic equation gives no useful information in our approximation, beyond that $\tau = t$. The other three components give the approximate equation of motion

$$
\ddot{\boldsymbol{r}} = -\tfrac{1}{2} \nabla(h_{00}),
\tag{6.5}
$$

where the dot can be differentiation with respect to either t or τ. Thus if we want to reduce general relativity to the Newtonian theory in this limit, then we must take $\phi = \tfrac{1}{2} h_{00}$, to within an added constant.

The choice is consistent with eqn (1.2), which with $\rho = 0$ is the approximate form of $R_{00} = 0$. To see this, we note that the derivatives of h_{ab} with respect to x^0 in (6.4) are all ignored, giving

$$R_{0b0}{}^d = \tfrac{1}{2} m^{de} \partial_b \partial_e h_{00}$$

in our approximation. Therefore the 00-component of the vacuum equation $R_{ab} = 0$ is

$$R_{0b0}{}^b = -\tfrac{1}{2}(\partial_1^2 h_{00} + \partial_2^2 h_{00} + \partial_3^2 h_{00}) = -\nabla^2 \phi = 0\,,$$

which is Laplace's equation, the vacuum equation in Newtonian theory. In this limit, Einstein's theory reduces to Newton's.

Exercise 6.1

What about the other nine components of Einstein's vacuum equation?

6.3 The Nonvacuum Case

What happens when there is matter present? What is the analogue of Poisson's equation $\nabla^2 \phi = 4\pi G \rho$?

We consider first the case in which the matter generating the gravitational field is a dust cloud. Its energy density is encoded in the energy-momentum tensor

$$T^{ab} = \rho U^a U^b\,,$$

where U^a is the four-velocity field of the dust and ρ is the energy (mass) density measured in the local rest frame. We know that $\partial_a T^{ab} = 0$ in local inertial coordinates at an event because the continuity equation holds in special relativity. Therefore in general coordinates we have

$$\nabla_a T^{ab} = 0\,,$$

because this is a tensor equation which reduces to $\partial_a T^{ab} = 0$ in local inertial coordinates.

The identification of R_{00} with $-\nabla^2 \phi$ suggests that the field equation in general relativity should equate R_{ab} to a constant multiple of T_{ab}. Unfortunately, this will not do because in general $\nabla_a R^{ab} \neq 0$. But there is a tensor closely related to the Ricci tensor which can be put on the left-hand side without contradiction. This is the *Einstein tensor*

$$G_{ab} = R_{ab} - \tfrac{1}{2} R g_{ab}\,,$$

where $R = R_a{}^a$ is the *Ricci scalar* or *scalar curvature*.

Proposition 6.1

For any space–time metric, $\nabla_a G^{ab} = 0$.

Proof

The Bianchi identity is

$$\nabla_a R_{bcde} + \nabla_b R_{cade} + \nabla_c R_{abde} = 0.$$

By contracting with $g^{ad} g^{ce}$, we obtain

$$0 = 2\nabla^a R_{ab} - \nabla_b R = 2\nabla^a (R_{ab} - \tfrac{1}{2} g_{ab} R) = 2\nabla^a G_{ab},$$

which completes the proof.   □

Our candidate for the field equation is $G_{ab} = k\rho U_a U_b$, with k constant. By contracting with g^{ab}, we obtain

$$R - 2R = kT_a{}^a = k\rho$$

because $g^{ab} g_{ab} = 4$ and $U^a U_a = 1$. So an equivalent form of the equation is

$$R_{ab} = k(T_{ab} - \tfrac{1}{2}\rho g_{ab}).$$

Now in the coordinates we used in the weak field limit, $R_{00} = -\nabla^2 \phi$, and $T_{00} = \rho$. Thus in this limit, we have $\nabla^2 \phi = -\tfrac{1}{2} k\rho$. To obtain the correct correspondence with the Newtonian theory, therefore, we must take $k = -8\pi G$, which means that the field equation is

$$R_{ab} - \tfrac{1}{2} R g_{ab} = -8\pi G T_{ab}.$$

Einstein proposed that this holds in general, with T_{ab} the sum of the energy-momentum tensors of all the matter present, including electromagnetic and other fields.

From now on, we always use units in which $G = 1 = c$. Given the unit of time, say the second, the condition $c = 1$ fixes the unit of distance, the light-second, and the normalization $G = 1$ fixes the unit of mass. With these choices, time, distance, and mass all have the same units.

EXERCISES

6.2. Calculate your age, height, and mass in seconds. Find the conversion factors to SI units and take note that our units are not likely to be useful for everyday purposes.

7
Spherical Symmetry

In this chapter, we find the gravitational field outside a spherical body of mass m. That is, we find the solution of the vacuum equation analogous to the Newtonian potential

$$\phi = -Gm/r\,.$$

Our derivation is in the form of an extended worked example, and can only be described as a 'head-on' approach. There are certainly more elegant ways of proceeding, but they require deeper knowledge of the theory. It is in any case instructive to see how complex is the direct solution of Einstein's equations even in this, the simplest nontrivial example. From this point on, we work in units in which $G = 1$.

7.1 The Field of a Static Spherical Body

By saying the body has mass m, we mean that the metric approaches that of Minkowski space for large r and that

$$g_{00} \sim 1 - 2m/r\,.$$

A long way from the body, the field is that of a static spherically symmetric body of mass m in the weak field limit. In operational terms, m is the mass measured by analysing orbits in the field of the body near infinity.

We want the metric to have the symmetries appropriate to a static spherical body. In spherical polar coordinates, the Minkowski space metric is

$$\mathrm{d}t^2 - \mathrm{d}r^2 - r^2(\mathrm{d}\theta^2 + \sin^2\theta\mathrm{d}\varphi^2)\,. \tag{7.1}$$

The expression in brackets is the metric on the unit sphere.

Our space–time metric must reduce to (7.1) when $m = 0$ and in any case in the limit $r \to \infty$. The flat metric (7.1) has the following features.

– The metric coefficients have no t-dependence.

– There are no $\mathrm{d}t\,\mathrm{d}r$, $\mathrm{d}t\,\mathrm{d}\varphi$, or $\mathrm{d}t\,\mathrm{d}\theta$ terms. It is therefore *time reversible*, or in other words, invariant under $t \mapsto -t$.

– There are no $\mathrm{d}r\,\mathrm{d}\theta$ or $\mathrm{d}r\,\mathrm{d}\varphi$ terms. At constant time, the radial vector is perpendicular to the surfaces of constant r.

– The metric on each surface of constant t and r is a constant multiple of the metric on the unit sphere.

– The coefficients of $\mathrm{d}t^2$ and $\mathrm{d}r^2$ are independent of θ and φ.

The first two characterize the flat metric as 'static'; the last three are what we mean by 'spherical symmetry'. We assume that our curved space–time metric has all these properties, and thus that it is of the form

$$\mathrm{d}s^2 = A(r)\mathrm{d}t^2 - B(r)\mathrm{d}r^2 - C(r)r^2(\mathrm{d}\theta^2 + \sin^2\theta\,\mathrm{d}\varphi^2)\,,$$

for some functions A, B, C of r.

There is no loss of generality in taking $C = 1$ because we are free to replace r by $r\sqrt{C}$. So our task is to solve the Einstein vacuum equation with $C = 1$, and with A, B subject to the boundary conditions $A, B \to 1$ and

$$A = 1 - 2m/r + O(r^{-2})$$

as $r \to \infty$.

7.2 The Curvature Tensor

We need to find R_{ab} in terms of A and B, with $C = 1$. The first step is to find the Christoffel symbols from the geodesic equations. These are the Lagrange equations

$$\frac{\mathrm{d}}{\mathrm{d}\tau}\left(\frac{\partial L}{\partial \dot{x}^a}\right) - \frac{\partial L}{\partial x^a} = 0 \tag{7.2}$$

of the Lagrangian

$$L = \tfrac{1}{2}\left(A\dot{t}^2 - B\dot{r}^2 - r^2\dot{\theta}^2 - r^2\sin^2\theta\,\dot{\varphi}^2\right),$$

with $x^0 = t$, $x^1 = r$, $x^2 = \theta$, $x^3 = \varphi$. The idea is to read off the Christoffel symbols by comparing (7.2) with the geodesic equations

$$\ddot{x}^a + \Gamma^a_{bc}\dot{x}^b\dot{x}^c = 0\,.$$

Written out in full, the Lagrange equations are

$$\frac{\mathrm{d}}{\mathrm{d}\tau}\left(A\dot{t}\right) = 0$$

$$\frac{\mathrm{d}}{\mathrm{d}\tau}\left(-B\dot{r}\right) - \tfrac{1}{2}A'\dot{t}^2 + \tfrac{1}{2}B'\dot{r}^2 + r\dot{\theta}^2 + r\sin^2\theta\dot{\varphi}^2 = 0$$

$$\frac{\mathrm{d}}{\mathrm{d}\tau}\left(-r^2\dot{\theta}\right) + r^2\sin\theta\cos\theta\,\dot{\varphi}^2 = 0$$

$$\frac{\mathrm{d}}{\mathrm{d}\tau}\left(-r^2\sin^2\theta\dot{\varphi}\right) = 0\,,$$

where the dot denotes the derivative with respect to τ. These can be rearranged as:

$$\ddot{t} + A'A^{-1}\dot{t}\dot{r} = 0$$

$$\ddot{r} + \tfrac{1}{2}A'B^{-1}\dot{t}^2 + \tfrac{1}{2}B'B^{-1}\dot{r}^2 - B^{-1}r\dot{\theta}^2 - B^{-1}r\sin^2\theta\,\dot{\varphi}^2 = 0$$

$$\ddot{\theta} + 2r^{-1}\dot{\theta}\dot{r} - \sin\theta\cos\theta\,\dot{\varphi}^2 = 0$$

$$\ddot{\varphi} + 2r^{-1}\dot{\varphi}\dot{r} + 2\cot\theta\,\dot{\theta}\dot{\varphi} = 0\,.$$

We can then read off the Christoffel symbols Γ^a_{bc} as

$(a = 0)$ $\Gamma^0_{01} = \Gamma^0_{10} = A'/2A$

$(a = 1)$ $\Gamma^1_{00} = A'/2B$, $\Gamma^1_{11} = B'/2B$, $\Gamma^1_{22} = -r/B$, $\Gamma^1_{33} = -r\sin^2\theta/B$

$(a = 2)$ $\Gamma^2_{21} = \Gamma^2_{12} = r^{-1}$, $\Gamma^2_{33} = -\sin\theta\cos\theta$

$(a = 3)$ $\Gamma^3_{31} = \Gamma^3_{13} = r^{-1}$, $\Gamma^3_{23} = \Gamma^3_{32} = \cot\theta\,.$

All the others vanish. Note carefully the factors of $\tfrac{1}{2}$ when $b \neq c$. Why are they there?

From the definition of the curvature tensor, we have

$$R_{abc}{}^d = \partial_a\Gamma^d_{bc} - \partial_b\Gamma^d_{ac} - \Gamma^d_{be}\Gamma^e_{ac} + \Gamma^d_{ae}\Gamma^e_{bc}\,.$$

The components R_{ac} of the Ricci tensor are then given by putting $b = d$ and summing. Thus

$$R_{00} = R_{010}{}^1 + R_{020}{}^2 + R_{030}{}^3$$

and so on. We find

$$
\begin{aligned}
R_{232}{}^{3} &= \partial_2 \Gamma^3_{32} - \partial_3 \Gamma^3_{22} - \Gamma^3_{3e}\Gamma^e_{22} + \Gamma^3_{2e}\Gamma^e_{32} \\
&= \partial_\theta(\cot\theta) + B^{-1} + \cot^2\theta \\
&= -1 + B^{-1} \\
R_{010}{}^{1} &= -A''/2B + B'A'/4B^2 + A'^2/4BA \\
R_{020}{}^{2} &= R_{030}{}^{3} = -A'/2Br \\
R_{121}{}^{2} &= R_{131}{}^{3} = -B'/2Br \\
R_{101}{}^{0} &= -BR_{010}{}^{1}/A \\
R_{303}{}^{0} &= -r^2\sin^2\theta R_{030}{}^{3}/A = r\sin^2\theta A'/2BA \\
R_{313}{}^{1} &= r^2\sin^2\theta R_{131}{}^{3}/B = -r\sin^2\theta B'/2B^2 \,.
\end{aligned}
$$

Hence the the vacuum equations are

$$
\begin{aligned}
R_{00} &= -A''/2B + B'A'/4B^2 + A'^2/4BA - A'/Br = 0 & (7.3) \\
R_{11} &= A''/2A - A'^2/4A^2 - B'A'/4BA - B'/Br = 0 & (7.4) \\
R_{22} &= R_{33}/\sin^2\theta = rA'/2BA - rB'/2B^2 + 1/B - 1 = 0 \,. & (7.5)
\end{aligned}
$$

All the other components of the Ricci tensor vanish identically, as can be seen by direct calculation or by using the fact that R_{ab} must have the same symmetries as the metric.

In all, we have three equations in the two unknowns A, B. Fortunately they are consistent. If we take B times (7.3) and add A times (7.4), then we get

$$
AB' + BA' = 0 \,,
$$

and hence that AB is constant. Because we want $A, B \to 1$ as $r \to \infty$, we must therefore have $AB = 1$. By substituting into (7.5), we then get that $rA' + A = 1$ and hence that

$$
A = \frac{1}{B} = 1 + \frac{k}{r}
$$

for some constant k. But for large r, we want $A = 1 - 2m/r + O(r^2)$, so $k = -2m$, and the solution is

$$
\mathrm{d}s^2 = \left(1 - \frac{2m}{r}\right)\mathrm{d}t^2 - \frac{\mathrm{d}r^2}{1 - 2m/r} - r^2(\mathrm{d}\theta^2 + \sin^2\theta\,\mathrm{d}\varphi^2) \,. \qquad (7.6)
$$

This is the *Schwarzschild metric*. The method of derivation is notable only for the incentive it gives to find more subtle methods for tackling Einstein's equations.

7.3 Stationary Observers

An observer in a fixed location relative to our coordinate system has a worldline with constant r, θ, φ, and therefore has four-velocity U with only the first component nonzero. Because $U^a U_a = 1$ and $U^0 > 0$, the four-velocity components are

$$U^0 = \frac{1}{\sqrt{1 - 2m/r}}, \qquad U^a = 0 \quad \text{for } a = 1, 2, 3.$$

The observer's worldline is not geodesic, as we know, for example, from the fact that an observer at rest on the earth's surface is accelerating relative to the local inertial frame and is not in free-fall. The observer interprets this acceleration as the 'force of gravity'.

In local inertial coordinates at an event, the four-acceleration is $\alpha^a = \mathrm{d}U^a/\mathrm{d}\tau$. In general coordinates, therefore,

$$\alpha^a = U^b \nabla_b U^a = U^b \partial_b U^a + \Gamma^a_{bc} U^b U^c.$$

As in special relativity, the acceleration actually felt by the observer is $\sqrt{-\alpha_a \alpha^a}$.

By using the Christoffel symbols found above, we have

$$\alpha^a = U^0 \partial_0 U^a + \Gamma^a_{00} U^0 U^0.$$

The only nonvanishing component is

$$\alpha^1 = \frac{A'(U^0)^2}{2B} = \tfrac{1}{2} A',$$

where $A = B^{-1} = 1 - 2m/r$. Thus the four-acceleration of the observer has components $(0, m/r^2, 0, 0)$, as one might expect by naive analogy with Newtonian theory. However, the acceleration felt by the observer is

$$g = \sqrt{-\alpha^a \alpha_a} = \frac{m}{r^2} \frac{1}{\sqrt{1 - 2m/r}}. \tag{7.7}$$

Thus the 'force of gravity' is given by the same inverse square law $g = m/r^2$ as in Newtonian theory for large r, but increases to infinity as r approaches the Schwarzschild radius $r = 2m$. We show later that $r = 2m$ is the event horizon of a black hole. What we are observing here is a consequence of the fact that inside a black hole one would have to travel faster than light in order to stay 'in the same place'.

7.4 Potential Energy

The worldlines of particles in free-fall and of photons in a general space–time are geodesics. They are the solutions of the differential equations generated by the Lagrangian

$$L = \tfrac{1}{2} g_{ab} \dot{x}^a \dot{x}^b \, .$$

In the case of free particles, the dot denotes differentiation with respect to *proper time* τ, the time measured by a clock carried by the particle. In the case of photons, the dot is differentiation with respect to an affine parameter, which is defined only up to a constant factor and the addition of a further constant.

In the Schwarzschild metric, the geodesic Lagrangian is

$$L = \frac{1}{2} \left[\left(1 - \frac{2m}{r} \right) \dot{t}^2 - \frac{\dot{r}^2}{1 - 2m/r} - r^2 \left(\dot{\theta}^2 + \sin^2 \theta \dot{\varphi}^2 \right) \right] \, . \qquad (7.8)$$

Therefore the t equation for the geodesic motion of a free particle is

$$\frac{\mathrm{d}}{\mathrm{d}\tau} \left(\frac{\partial L}{\partial \dot{t}} \right) = 0$$

because the Lagrangian is independent of t. Consequently

$$E = (1 - 2m/r)\dot{t}$$

is constant along the particle worldline. What is the interpretation of this constant? Suppose that the particle has four-velocity V and unit mass. Then relative to an observer 'at rest' at some point in the particle's history, the particle has speed v given by

$$\gamma(v) = \frac{1}{\sqrt{1 - v^2}} = U^a V_a = g_{00} U^0 V^0 = \dot{t}\sqrt{1 - 2m/r} \, .$$

Therefore

$$E = \frac{\sqrt{1 - 2m/r}}{\sqrt{1 - v^2}} \, .$$

For large r and small v, this is approximately

$$E = 1 + \tfrac{1}{2} v^2 - m/r + \text{smaller terms} \, .$$

Thus E is the sum of the rest energy (Mc^2 with $M = 1$ and $c = 1$), the kinetic energy $\tfrac{1}{2} v^2$ relative to the observer, and the Newtonian potential energy $-m/r$. Thus it is reasonable to interpret E as the *total energy* of the particle. We note that this is consistent with (7.7), which can be written

$$g = \partial_r \sqrt{1 - 2m/r} \, ,$$

with the implication that we should interpret $\sqrt{1 - 2m/r}$ as the potential energy of a unit mass particle at rest. Conservation of energy is then a consequence of $\partial L/\partial t = 0$, that is, of the fact that t is an ignorable coordinate. As in classical mechanics, energy is conserved when there is invariance under time translation.

7.5 Photons and Gravitational Redshift

In special relativity, a photon worldline is a null line. The *frequency four-vector* K is tangent to the worldline and encodes information about the frequency of the photon, as measured by a moving observer. If the observer has four-velocity U, then the observed frequency is $\omega = U_a K^a$. The frequency four-vector is constant along the photon worldline.

By our usual principle that special relativity should hold over short times and distances in local inertial coordinates, it follows that in general relativity K is tangent to the photon worldline, which is now a null geodesic, and that

$$K^a \nabla_a K^b = 0 \, .$$

If we put $W^a = \mathrm{d}x^a/\mathrm{d}\sigma$, where σ is the affine parameter, then the geodesic equation is

$$W^a \nabla_a W^b = 0 \, .$$

Because W is proportional to K and because it is tangent to the photon worldline, it must in fact be a constant multiple of K. By rescaling σ, we can take $W^a = K^a$. With this choice of σ, the frequency four-vector is given by

$$K^a = \mathrm{d}x^a/\mathrm{d}\sigma \, .$$

Now consider two observers O_1 and O_2 in the Schwarzschild space–time, at rest relative to the Schwarzschild coordinates at $r = r_1$ and $r = r_2$, respectively. If O_1 sends out a photon to O_2, and if the frequency measured by O_1 at transmission is ω_1, then what is the frequency at reception as measured by O_2?

Denote the photon's worldline by $x^a = x^a(\sigma)$, where the affine parameter σ is chosen so that the frequency four-vector is $K^a = \mathrm{d}x^a/\mathrm{d}\sigma$. Let ω denote frequency measured by a stationary observer at r. Then we have

$$\omega = U^a K_a = g_{00} U^0 K^0 = \dot{t}\sqrt{1 - 2m/r} \, ,$$

where the dot is the derivative with respect to σ. However $(1 - 2m/r)\dot{t}$ is constant along the worldline because L is independent of t. Therefore $\omega\sqrt{1 - 2m/r}$

is also constant and so we have

$$\omega_2 = \omega_1 \sqrt{\frac{1 - 2m/r_1}{1 - 2m/r_2}} \, .$$

This is the *gravitational redshift* formula. For large r_1, r_2, we have

$$\omega_2 \sim \omega_1 (1 + m/r_2 - m/r_1) \, ,$$

so the change in frequency is proportional to the difference gravitational potential between the two observers. This is precisely what is needed to avoid the paradox in Bondi's perpetual motion machine. We remark also that quantum theory tells us that the energy of a photon relative to an observer is $\hbar\omega$. So the conservation law here can again be interpreted as 'conservation of energy'.

7.6 Killing Vectors

A special role is played in these calculations by time symmetry. It is this that allows us to say what we mean by 'stationary' observers, and it is this that gives us energy conservation.

More generally, if the metric coefficients g_{ab} are independent of one of the coordinates x^0, then $L = \frac{1}{2} g_{ab} \dot{x}^a \dot{x}^b$ is independent of x^0, and so from Lagrange's equations

$$\frac{\partial L}{\partial \dot{x}^0} = g_{a0} \dot{x}^a$$

is constant along geodesics. But this quantity is equal to $T^a V_a$, where $V^a = \dot{x}^a$ and T is the four-vector field with components $(1, 0, 0, 0)$. The quantity $T^a V_a$ is an invariant. It depends only on the four-vectors V and T, and not on the choice of coordinates, although, of course, T will have components $(1, 0, 0, 0)$ only for particular choices of coordinates.

Definition 7.1 (Preliminary definition)

A nonvanishing vector field T is said to be a *Killing vector field* or *Killing vector* whenever there exists a coordinate system in which T has components $(1, 0, 0, 0)$ and g_{ab} is independent of x^0.

We have just proved the following.

Proposition 7.2

If T is a Killing vector, then $T_a \dot{x}^a$ is constant along any geodesic.

How can we recognise a Killing vector, and therefore derive a conserved quantity for free particle and photon orbits, without making the transformation to the special coordinate system? To answer this, we look first at the defining property in the special coordinates in which T has components $(1, 0, 0, 0)$. Here we have

$$0 = \partial_0 g_{ab} = T^c \partial_c g_{ab}\,.$$

But we also have

$$\nabla_a T_b = \partial_a(g_{bc}T^c) - \tfrac{1}{2}T^c(\partial_a g_{bc} + \partial_b g_{ac} - \partial_c g_{ab})$$
$$\nabla_b T_a = \partial_b(g_{ac}T^c) - \tfrac{1}{2}T^c(\partial_b g_{ac} + \partial_a g_{bc} - \partial_c g_{ba})\,.$$

By adding, we get

$$\nabla_a T_b + \nabla_b T_a = T^c \partial_c g_{ab} = 0$$

because $\partial_a T^c = 0$. But the left-hand side is a tensor. Therefore it vanishes in one coordinate system if and only if it vanishes in every coordinate system. We have proved the following.

Proposition 7.3

Let T^a be a nonvanishing vector field. If T^a is a Killing vector then

$$\nabla_a T_b + \nabla_b T_a = 0$$

in any coordinate system.

The converse is also true. The proof relies on the fact that for any nonvanishing vector field $T \neq 0$, there exists a local coordinate system in which T has components $(1, 0, 0, 0)$. One deduces the proposition by working in such coordinates and by following through the same calculation in reverse.

We can use Proposition 7.3 to prove Proposition 7.2 directly by starting from the geodesic condition in the form $V^a \nabla_a V_b = 0$, where $V^a = \dot{x}^a$. From this we get that the derivative of $V^a T_a$ is

$$V^a \nabla_a(V^b T_b) = V^a V^b \nabla_a T_b = \tfrac{1}{2}V^a V^b(\nabla_a T_b + \nabla_b T_a) = 0\,,$$

and hence that $V^a T_a$ is constant. The converse statement can also be deduced from this. If $\dot{x}^a T_a$ is conserved along every geodesic, then T_a is a Killing vector.

We use Proposition 7.3 to extend the definition by dropping the condition that T^a should be everywhere nonvanishing. It now takes the following form.

Definition 7.4 (Standard definition)

A vector field T^a is a *Killing vector* if $\nabla_a T_b + \nabla_b T_a = 0$.

EXERCISES

7.1. A clock is said to be *at rest* in the Schwarzschild space–time if its r, θ, and φ coordinates are constant. Show that the coordinate time and the proper time along the clock's worldline, that is, the time τ shown on the clock, are related by

$$\frac{\mathrm{d}t}{\mathrm{d}\tau} = \left(1 - \frac{2m}{r}\right)^{-1/2}.$$

Note that the worldline is not a geodesic.

Show that along a radial null geodesic, that is, one on which only t and r are varying,

$$\frac{\mathrm{d}t}{\mathrm{d}r} = \frac{r}{r - 2m}.$$

Two clocks C_1 and C_2 are at rest at (r_1, θ, φ) and (r_2, θ, φ). A photon is emitted from C_1 at event A and arrives at C_2 at event B. A second photon is emitted from C_1 at event A' and arrives at C_2 at event B'. Show that the coordinate time interval Δt between A and A' is the same as the coordinate time interval between B and B'. Hence show that the time interval $\Delta \tau_1$ between A and A' measured by C_1 is related to the time interval $\Delta \tau_2$ between B and B' measured by C_2 by

$$\Delta \tau_1 \left(1 - \frac{2m}{r_1}\right)^{-1/2} = \Delta \tau_2 \left(1 - \frac{2m}{r_2}\right)^{-1/2}.$$

If you wear two watches, one on your wrist and one on your ankle, and you synchronize them at the beginning of the year, by how much is the watch on your wrist faster or slower than the one on your ankle at the end of a year? (Assume that you spend the whole year standing upright without moving. In general units, you must replace m/r by Gm/rc^2.)

7.2. Show that if X is a vector field and T_{ab} is a tensor field of type $(0, 2)$, then

$$X^a \partial_a T_{bc} + T_{ac} \partial_b X^a + T_{ba} \partial_c X^a$$

transforms as a tensor of type (0,2). This tensor is called the *Lie derivative* of T along X.

Show that X is a Killing vector if and only if the Lie derivative of the metric along X vanishes. Show that if X and Y are Killing vectors, then so is the vector field $[X, Y]$, which is defined by

$$[X, Y]^a = X^b \partial_b Y^a - Y^b \partial_b X^a \,.$$

Let g_{ab} be the Schwarzschild metric, with $x^0 = t$, $x^1 = r$, $x^2 = \theta$, $x^3 = \varphi$. Show that the following are the components of Killing vectors

$$(1, 0, 0, 0), \quad (0, 0, 0, 1), \quad (0, 0, -\cos\varphi, \cot\theta \sin\varphi)$$

and find a fourth Killing vector which is not a linear combination with constant coefficients of these three.

7.3. Show that if $B_a = \nabla_a f$ for some function f, then $\nabla_{[a} B_{b]} = 0$. The converse is also true (locally), and you may use this without proof.

Let F_{ab} be a solution of Maxwell's equations $\nabla_a F^{ab} = 0$, $\nabla_{[a} F_{bc]} = 0$ in curved space–time. The equation of motion of a particle of charge e and rest mass M is given by the Lorentz equation

$$M u^b \nabla_b U^a = e F^{ab} U_b,$$

where $U^a = \mathrm{d}x^a / \mathrm{d}\tau$, with τ the proper time. Show that if the Lie derivative of F_{ab} along X^a vanishes (see the previous exercise), then $F_{ab} X^b = \nabla_a f$ for some function f. Show that if X^a is also a Killing vector then $M u^a X_a + e f$ is a constant of motion for the particle.

Orbits in the Schwarzschild Space–Time

We now look at particle motion in the Schwarzschild background. Our main aim is to derive corrections to Kepler's laws, so we think of the gravitational field as that of the sun. By a 'particle', we mean a very small body, such as a planet, whose own gravitational field can be ignored.

8.1 Massive Particles

The particle orbits are generated by the Lagrangian

$$L = \frac{1}{2}\left[\left(1 - \frac{2m}{r}\right)\dot{t}^2 - \frac{\dot{r}^2}{1 - 2m/r} - r^2\left(\dot{\theta}^2 + \sin^2\theta\dot{\varphi}^2\right)\right],$$

where the parameter is the proper time τ and the dot is the derivative with respect to τ. We assume that $r > 2m$, which means that we are looking at the external field of a spherical star rather than the field inside a black hole. Because L has no explicit dependence on t, φ, or τ, we have three conservation laws.

$(\partial_t L = 0)$ $\quad E = (1 - 2m/r)\dot{t} = \text{constant}$
$(\partial_\varphi L = 0)$ $\quad J = r^2\sin^2\theta\,\dot{\varphi} = \text{constant}$
$(\partial_\tau L = 0)$ $\quad L = \text{constant}.$

In fact $g_{ab}\dot{x}^a\dot{x}^b = 1$ because τ is proper time, and so the third conservation law is simply $L = \frac{1}{2}$.

We need one other equation to determine the orbits. We use the θ Lagrange equation,

$$\frac{\mathrm{d}}{\mathrm{d}\tau}\left(r^2\dot{\theta}\right) - r^2\sin\theta\cos\theta\,\dot{\varphi}^2 = 0\,. \tag{8.1}$$

We could also write down the r equation, but it would contain no new information because, with the conservation laws, we already have four equations for the four unknown coordinates t, r, θ, φ.

Equation (8.1) is symmetric under

$$\theta \mapsto \pi - \theta\,.$$

Therefore an orbit on which $\theta = \pi/2$, $\dot{\theta} = 0$ at $\tau = 0$ will have $\theta = \pi/2$ for all τ. Because the field is spherically symmetric, we can understand all the orbits by studying only these equatorial orbits. There is no loss of generality, therefore, in putting $\theta = \pi/2$. We then have

$$1 = \frac{E^2}{1 - 2m/r} - \frac{\dot{r}^2}{1 - 2m/r} - \frac{J^2}{r^2}$$

by combining the conservation laws. That is,

$$\dot{r}^2 = -\frac{J^2}{r^2}\left(1 - \frac{2m}{r}\right) + E^2 - \left(1 - \frac{2m}{r}\right)\,.$$

This is a first-order differential equation for r as a function of proper time. As in Newtonian theory, the equation looks a bit simpler if we replace r by $u = m/r$ and use φ instead of τ as the parameter. Now

$$\frac{\mathrm{d}u}{\mathrm{d}\varphi} = -\frac{m}{r^2}\frac{\mathrm{d}r}{\mathrm{d}\tau}\bigg/\frac{\mathrm{d}\varphi}{\mathrm{d}\tau} = -\frac{m\dot{r}}{J}\,.$$

Therefore the orbits are given by

$$\left(\frac{\mathrm{d}u}{\mathrm{d}\varphi}\right)^2 = \frac{m^2E^2}{J^2} - u^2(1 - 2u) - \frac{m^2(1 - 2u)}{J^2}\,, \tag{8.2}$$

provided that $J \neq 0$, that is, provided that the orbit is not radial.

8.2 Comparison with the Newtonian Theory

In the corresponding problem in Newton's theory, the particle (assumed to have unit mass) moves under the influence of the inverse square law force m/r^2. The equatorial orbits are determined in plane polar coordinates r, φ by the conservation of the energy ε and the angular momentum J, by

$$\varepsilon = \tfrac{1}{2}(\dot{r}^2 + r^2 \dot{\varphi}^2) - m/r, \qquad J = r^2 \dot{\varphi}.$$

As in the Schwarzschild space–time, we put $u = m/r$, $du/d\varphi = -m\dot{r}/J$. Then we have

$$\varepsilon = \frac{J^2}{2m^2} \left(\frac{du}{d\varphi} \right)^2 + \frac{J^2 u^2}{2m^2} - u.$$

To make comparison between the two theories, we put $\beta = m/J$, and $p = du/d\varphi$. In the Newtonian case, we put $k = \varepsilon m^2/J^2$ and define

$$g(u) = 2\beta^2 u + 2k - u^2.$$

In general relativity, we put $k = (E^2 - 1)m^2/2J^2$ and define

$$f(u) = 2\beta^2 u + 2k - u^2 + 2u^3.$$

Then the orbits are given by $p^2 = g(u)$ in Newtonian theory and by $p^2 = f(u)$ in general relativity. The only difference is the extra term $2u^3$ in $f(u)$, which, of course, is small when r is large. In both cases, we are working in units in which $G = 1$.

The differential equations for the orbits can also be written in the second-order form

$$\frac{d^2 u}{d\varphi^2} = \tfrac{1}{2} f'(u)$$

in general relativity, or in the same way with $\tfrac{1}{2} g'(u)$ on the right in Newtonian theory.

We can see the effect of the extra term in one of the classic tests of general relativity, the perihelion advance of Mercury. In general relativity, the point on a planet's orbit at which it is closest to the sun—the perihelion—advances on each orbit. In Newtonian theory the orbit is closed and the perihelion is always in the same position, provided that one ignores the effect of other planets. The relativistic advance is most significant in the case of Mercury because its orbit is closest to the sun, where the sun's field is strongest. In fact the perihelion also advances in Newtonian theory because of interactions with other planets, most notably with Jupiter. The general relativistic effect is the additional advance that cannot be explained in this way. When it was first observed, Le Verrier suggested that the additional advance might be due to another planet with an orbit closer to the sun than Mercury's. He predicted that it would be visible crossing the sun's disc in March 1877, but it was not seen [1].

8.3 Newtonian Orbits

In the Newtonian theory, we have

$$\frac{\mathrm{d}^2 u}{\mathrm{d}\varphi^2} + u = \beta^2 \,, \tag{8.3}$$

which implies that

$$u = \beta^2 + A\cos(\varphi - \varphi_0) \,,$$

for constant A, φ_0. By differentiating we get

$$p^2 = A^2 \sin^2(\varphi - \varphi_0) = A^2 - (u - \beta^2)^2 \,.$$

Hence $A^2 = 2k + \beta^4$. The form of the orbit depends on the sign of k.

(1) If $k > 0$ then $|A| > \beta^2$ and $u = 0$ for some values of φ. In this case, the orbits are hyperbolic and the particle can escape to infinity.

(2) If $k < 0$ then $|A| < \beta^2$. In this case, u is bounded away from zero, and therefore $|r|$ is bounded and the orbits are elliptic.

A special case arises in (2) when u is constant on the orbit, and so

$$\frac{\mathrm{d}u}{\mathrm{d}\varphi} \qquad \text{and} \qquad \frac{\mathrm{d}^2 u}{\mathrm{d}\varphi^2}$$

vanish identically. Such circular orbits are given by solving

$$g(u_0) = g'(u_0) = 0$$

for the constant value u_0 of u. The result is $u_0 = \beta^2$, where

$$\beta^4 + 2k = 0 \,.$$

For a general orbit with $k < 0$, we can rewrite (8.3) in the form

$$\frac{\mathrm{d}^2 v}{\mathrm{d}\varphi^2} + v = 0 \,,$$

where $v = u - \beta^2$. This is the equation of simple harmonic motion with period 2π. The 'time' of course is not t, but the polar angle φ. Thus we can think of a general elliptic orbit as oscillating about a circular orbit ($u = \beta^2$) with simple harmonic motion. The fact that in these oscillations the period of u as a function of φ is exactly 2π is what makes the elliptic orbits closed in Newtonian theory. Each circuit of the origin adds 2π to φ and brings the particle back to the initial value of u. In particular, perihelion always occurs at the same value of φ. There is no perihelion advance in the two-body system.

One can gain some insight into the structure of the orbits in Newtonian theory by plotting the *phase portrait*, in which one represents the orbits by curves in the u, p-plane. If we fix β and plot the phase curves for varying values of k, the result is a set of concentric circles

$$p^2 + (u - \beta^2)^2 = 2k + \beta^4$$

centred on the circular orbit $u_0 = \beta^2$, labelled A in Figure 8.1. The hyperbolic orbits are those that meet the p-axis; the elliptic orbits are those that do not. The two families are separated by the parabolic orbit, which touches the p-axis at the origin.

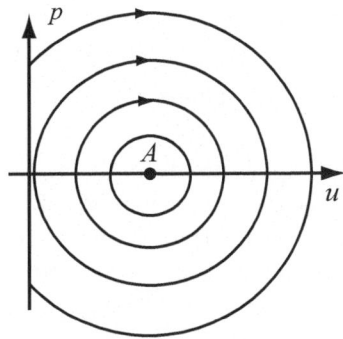

Figure 8.1 The Newtonian phase portrait

8.4 The Perihelion Advance

In general relativity, there are also closed orbits $u = u_0$. The corresponding values of the constants β and E are found by solving

$$f'(u_0) = 0 = f(u_0) \,.$$

Now consider an orbit $u = u_0 + v(\varphi)$ which is almost circular, so that v is small. By substituting into the equation of motion, we obtain

$$\frac{\mathrm{d}^2 v}{\mathrm{d}\varphi^2} = \tfrac{1}{2} f'(u) = \tfrac{1}{2} f'(u_0) + \tfrac{1}{2} v f''(u_0) + O(v^2) \,.$$

But $f'(u_0) = 0$, and $f''(u) = -2 + 12u$. Therefore v satisfies

$$\frac{d^2v}{d\varphi^2} + (1 - 6u_0)v = 0\,,$$

on ignoring the term $O(v^2)$. This is again the equation of simple harmonic motion, with φ as 'time'. So at least for orbits that are close to circular, we again have the picture that the planet's orbit oscillates about a circular orbit. Now, however, the period is not 2π, but

$$\varphi = \frac{2\pi}{\sqrt{1 - 6u_0}} \sim 2\pi + 6u_0\pi\,,$$

for small u_0, that is, for large r_0. Thus if the particle starts at perihelion where r minimal and u is maximal, then r returns to its initial value not after a whole rotation, but after φ has advanced through a further angle $6u_0\pi$. This is the *perihelion advance*. If we substitute $u_0 = m/r_0$ and put back in the constants—there is only one way to do this to get the dimensions right—then the advance is

$$\frac{6Gm\pi}{r_0c^2}$$

per revolution for an orbit of approximate radius r_0. We are ignoring second-order terms in $1/r_0$, as well as assuming that the orbit is 'nearly' circular.

In the case of the orbit of Mercury, the relevant quantities have the following values in SI units. The mass of the sun is $m = 1.98 \times 10^{30}$. The radius of the orbit is $r_0 = 5.79 \times 10^{10}$, and the constants are $G = 6.67 \times 10^{-11}$, and $c^2 = 9 \times 10^{16}$. This gives the advance as around $40''$ per century. A more careful analysis gives $43''$, exactly accounting for the anomaly without the need for Le Verrier's additional planet.

The effect is more marked in the case of the binary pulsar PSR 1913+16, where the advance is around $4°$ per year [21]. The system consists of a neutron star, about 15 miles across, but with a mass about 50% larger than that of the sun, orbiting another star once every 8 hours or so. Here one reverses the Mercury observation, using the rotation of the orbit to measure the masses. One then calculates the theoretical rate at which the orbital period should decrease as the two stars lose energy through gravitational radiation. The result, over 15 years, agrees with observation to within 0.5%

8.5 Circular Orbits

The equatorial orbits in the Schwarzschild space–time are given by $p^2 = f(u)$, where

$$f(u) = 2\beta^2 u - u^2 + 2u^3 + 2k\,,$$

and

$$u = \frac{m}{r}, \qquad p = \frac{\mathrm{d}u}{\mathrm{d}\varphi}, \qquad \beta = \frac{m}{J}, \qquad k = \frac{(E^2 - 1)m^2}{2J^2}.$$

They are solutions to the second-order equation

$$\frac{\mathrm{d}^2 u}{\mathrm{d}\varphi^2} = \tfrac{1}{2}f'(u).$$

The *circular* orbits are those for which r and therefore also u are constant. They are given by $f(u) = 0$, $f'(u) = 0$. The second of these equations implies that

$$6u = 1 \pm \sqrt{1 - 12\beta^2},$$

which for small β has solution

$$u = \beta^2 \quad \text{and} \quad u = \tfrac{1}{3} - \beta^2$$

on ignoring terms of order β^4. The first root is the Newtonian circular orbit. This is still present in general relativity provided that the radius m/β^2 is large compared to m. The second is a new feature. It has radius close to $r = 3m$, which is only just above the Schwarzschild radius $r = 2m$, and it exists only if the source of the gravitational field is contained within the sphere $r = 3m$, so the metric still takes the Schwarzschild form at this radius. We show below that $r = 3m$ itself is a circular photon orbit. A particle on the inner circular orbit has to be moving close to the velocity of light, relative to a stationary observer.

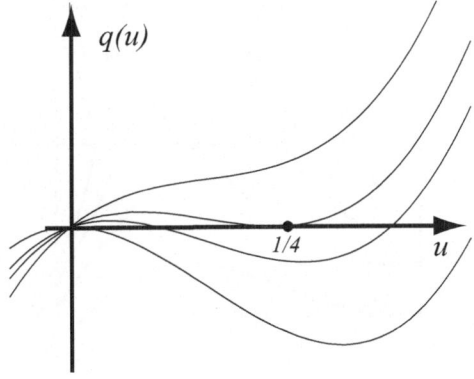

Figure 8.2 Plots of $q(u) = 2\beta^2 u - u^2 + 2u^3$

8.6 The Phase Portrait

We can understand more clearly the pattern of orbits by drawing the phase portrait in the u, p-plane for fixed β^2 and by varying k, as we did in the Newtonian theory. We first plot the graphs of $q(u) = u(2u^2 - u + 2\beta^2)$ in the q, u-plane for different values of β^2, in Figure 8.2. These curves in the q, u-plane have the following features.

- For $\beta^2 = 0$, the curve lies below the u-axis for $0 < u < \frac{1}{2}$, and touches it at the origin.

- For $\beta^2 = 1/16$ the two roots of $2u^2 - u + 2\beta^2$ come into coincidence.

- For $\beta^2 = 1/12$, the two roots of $q'(u)$ come into coincidence.

We consider the orbits only for $\frac{1}{2} > u > 0$, that is, for $r > 2m$. If the vacuum region extends that far inwards, the portion of space–time in which $r < 2m$ is inside a black hole. For each value of β^2, we get a phase portrait by plotting the curves $p^2 = q(u) + 2k$ for different values of k. For small u, that is, large r, the portrait coincides with the Newtonian picture. The differences arise as u approaches $\frac{1}{2}$. Note that the arrows in the plots show the direction of increasing φ, not of increasing time.

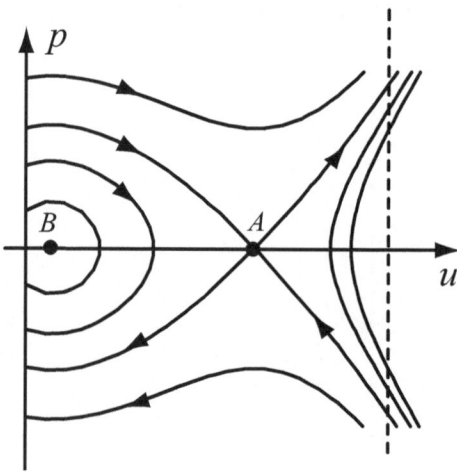

Figure 8.3   The case $0 < \beta^2 < 1/16$

The case $0 < \beta^2 < 1/16$

There are two circular orbits, one stable (B), and the other unstable with $k > 0$ (A). A particle disturbed from the inner circular orbit—the unstable one—can either spiral inwards or escape to infinity. The horizon is shown as a dashed line.

The case $1/16 < \beta^2 < 1/12$

The inner unstable circular orbit A has $k < 0$: a particle disturbed from this orbit will not escape to infinity. As β^2 is increased, the two circular orbits move towards each other. They coincide when β^2 reaches $1/12$, at $r = 6m$.

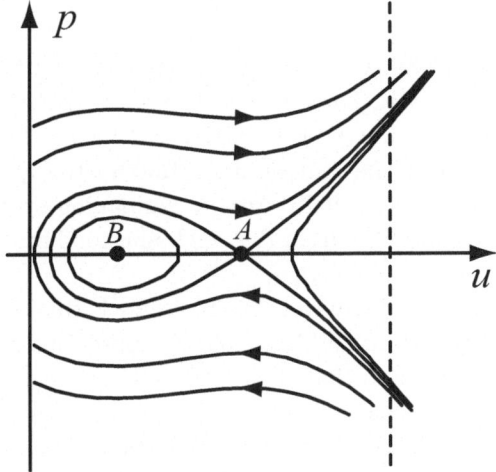

Figure 8.4 The case $1/16 < \beta^2 < 1/12$

The case $\beta^2 > 1/12$

There are no closed orbits in this case: the angular momentum is too small. All orbits either escape to infinity or spiral inwards.

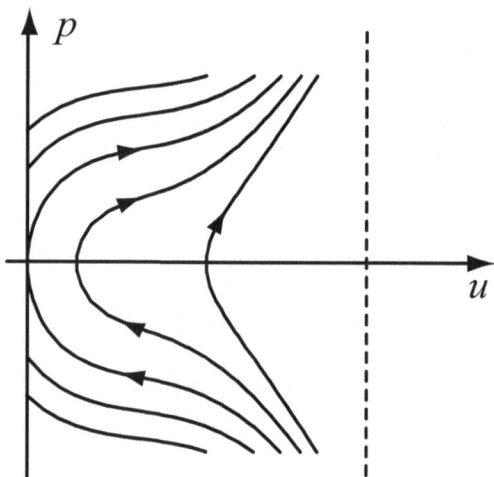

Figure 8.5 The case $\beta^2 > 1/12$

In no case are there stable circular orbits with $r < 6m$: this is the minimum radius for a planetary orbit. For a star of the mass of the sun, the minimum radius is 9 km. For an ordinary star, this is well inside the star itself, so the limit is not relevant. But the limit is important in the analysis of the infall of matter into a black hole, usually from a companion star. It is this that is responsible for X-ray emissions from the neighbourhood of a stellar mass black hole.

An interesting lesson to learn from the first case is that, contrary to popular belief, it is not easy to fall into a black hole. Suppose that initially the particle is at $r = r_0$, with $r_0 \gg r$, and that the radial and transverse components of its velocity relative to a stationary observer are v_r and v_t, respectively. As long as these are small compared with the velocity of light, we have that

$$E \sim 1 + \tfrac{1}{2}(v_t^2 + v_r^2) - m/r_0 = 1 + \epsilon, \qquad \beta = \frac{m}{J} = \frac{m}{r_0 v_t},$$

where ϵ is small. On the subsequent orbit, we must have

$$f(u) = \beta^2 u - u^2 + 2u^3 + \epsilon\beta^2 > 0.$$

If we ignore the last term in f, then the phase-plane analysis tells us that we must have $\beta^2 > 1/16$ on an orbit with $k \sim 0$ if the particle is to reach the horizon. That is,

$$v_t^2 < 16m^2/r_0^2.$$

Let $v_0 = \sqrt{m/r_0}$ denote the velocity of a circular orbit at the initial radius. Then the condition for our particle to fall into the black hole is

$$\left| \frac{v_t}{v_0} \right| < 4 \sqrt{\frac{m}{r_0}} = 4 \sqrt{\frac{R}{2r_0}} \,,$$

where R is the Schwarzschild radius. Although the black hole has a very strong gravitational field, its radius is small, and it presents an almost impossible target from any distance.

Exercise 8.1

What is $\sqrt{R/2r_0}$ if r_0 is the radius of the earth's orbit and the black hole has the mass of the sun?

8.7 Photon Orbits

The photon orbits are generated by the same Lagrangian

$$L = \frac{1}{2} \left[\left(1 - \frac{2m}{r} \right) \dot{t}^2 - \frac{\dot{r}^2}{1 - 2m/r} - r^2 \left(\dot{\theta}^2 + \sin^2 \theta \dot{\varphi}^2 \right) \right] ,$$

but now the dot is differentiation with respect to an affine parameter σ. Again we have three constants: the energy

$$E = (1 - 2m/r)\dot{t} \,,$$

the angular momentum

$$J = r^2 \sin^2 \theta \, \dot{\varphi} \,,$$

and the value L itself. In the case of photons, the last constant is zero because $g_{ab} \dot{x}^a \dot{x}^b$ vanishes when $\dot{x}^a$ is null.

By proceeding along the same lines as before, we obtain

$$p^2 = \alpha^2 + 2u^3 - u^2 \,, \tag{8.4}$$

where $u = m/r$, $p = du/d\varphi$, and $\alpha^2 = m^2 E^2/J^2$. There is no sensible Newtonian model with which to make comparisons, but we note that without the term $2u^3$, the orbits would be given by

$$u = \alpha \cos(\varphi - \varphi_0), \qquad p = -\alpha \sin(\varphi - \varphi_0) \,.$$

That is, by

$$r \cos(\varphi - \varphi_0) = m/\alpha \,, \tag{8.5}$$

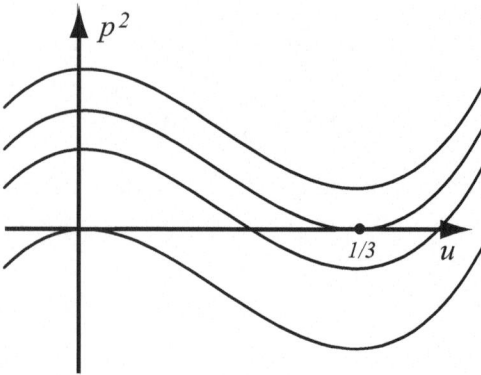

Figure 8.6 $\alpha^2 + 2u^3 - u^2$ against u for different values of α

which is the polar equation of a straight line. So we can think of the term $2u^3$ as the gravitational contribution. It is responsible for the 'bending of light by gravity'.

The phase portrait is found in the same way as before. First we plot

$$p^2 = \alpha^2 + 2u^3 - u^2$$

against u for different values of α, as in Figure 8.6. By taking the square root, we then get the phase portrait in Figure 8.7, with the different values of α labelling the different curves in the p, u-plane. Note that the arrows in the plot show the direction of increasing φ, not of increasing time. For large r (small u), the orbits look like hyperbolic Newtonian orbits. A photon travelling in from infinity will escape to infinity, but the trajectory will be deflected. As r decreases, the deflection increases, and the orbit can wind around the source many times. At $r = 3m$, it is possible for the photon to orbit in an unstable circular orbit.

8.8 The Bending of Light

We now find the approximate orbits for large r, that is, in the domain in which $u^2 \ll u^3$. They are approximately straight lines, but slightly bent by the gravitational field through the appearance of the 'gravitational term' $2u^3$ in (8.4).

To see the effect of the gravitational term, we put

$$u = \alpha \cos \varphi + v,$$

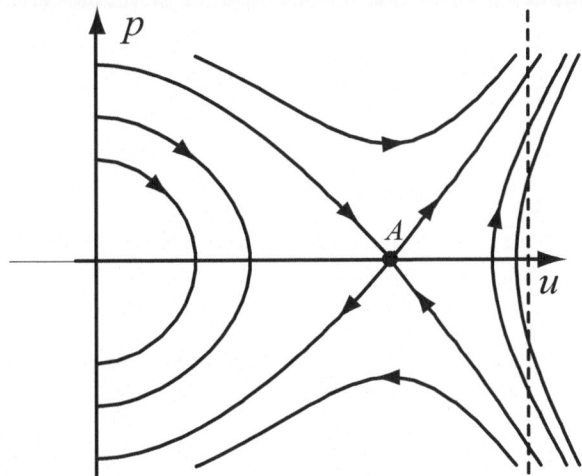

Figure 8.7 The phase portrait for photon orbits

where α is small and $v = O(\alpha^2)$. In other words, we consider a perturbation of (8.5) with $\varphi_0 = 0$. On ignoring terms of order $v^2 = O(\alpha^4)$, we get

$$
\begin{aligned}
0 &= u'^2 + u^2 - 2u^3 - \alpha^2 \\
&= -2\alpha v' \sin\varphi + 2\alpha v \cos\varphi - 2\alpha^3 \cos^3\varphi \,,
\end{aligned}
$$

where the prime is differentiation with respect to φ. From this we obtain

$$
\sin\varphi \, v' = v \cos\varphi - \alpha^2 \cos^3\varphi
$$

and hence on finding the integrating factor,

$$
\frac{\mathrm{d}}{\mathrm{d}\varphi}\left(\frac{v}{\sin\varphi}\right) = -\alpha^2 \cos\varphi \left(\frac{1}{\sin^2\varphi} - 1\right) .
$$

Therefore

$$
\frac{v}{\sin\varphi} = \frac{\alpha^2}{\sin\varphi} + \alpha^2 \sin\varphi + K
$$

for some constant K. We can set $K = 0$ by adjusting φ_0, to get

$$
v = \alpha^2(1 + \sin^2\varphi) \,.
$$

Because the gravitational field is weak, we can interpret r, φ as plane polar coordinates in the equatorial plane. The unperturbed trajectory is the straight line

$$
\alpha r \cos\varphi = m \,,
$$

which goes to infinity as $\varphi \to \pm\pi/2$. The perturbed trajectory, on the other hand, goes to infinity as

$$\varphi \to \pm(\pi/2 + \gamma) \,,$$

where the angle γ is given by

$$-\alpha \sin \gamma + \alpha^2 (1 + \cos^2 \gamma) = 0 \,.$$

This gives $\gamma = 2\alpha$ to the first order in α. So the total deflection of the light ray is

$$2\gamma = 4\alpha = 4m/D \,,$$

where $D = m/\alpha$ is the value of r at the point of closest approach to the source of the unperturbed trajectory. In SI units, the deflection is

$$\frac{4mG}{Dc^2} \,.$$

For a light ray just grazing the surface of the sun, we have (in SI units) $D = 7 \times 10^8$ (the radius of the sun), $m = 2 \times 10^{30}$ (the mass of the sun), $c = 3 \times 10^8$, and $G = 7 \times 10^{-11}$. The result is a deflection of 10^{-5} radians or $2''$. This is hard to observe because of the difficulty in detecting light that grazes the sun. Its effect can, however, be seen during a total eclipse, and was first observed by Eddington in 1919. The deflection causes the stars near the sun in the sky to appear to move from their normal positions away from the centre of the sun. Eddington compared photographs of the star field near the sun during a total eclipse with a photograph of the same star field when the sun was in a different position in the sky at another time of year [5].

EXERCISES

8.2. Show that along free particle worldlines in the equatorial plane of the Schwarzschild metric, the quantities

$$J = r^2 \dot{\varphi} \quad \text{and} \quad E = \left(1 - \frac{2m}{r}\right) \dot{t}$$

are constant. Here the dot is the derivative with respect to proper time. Explain why the particle cannot escape to infinity if $E < 1$.

Show that

$$\dot{r}^2 + \left(1 + \frac{J^2}{r^2}\right)\left(1 - \frac{2m}{r}\right) = E^2,$$

$$\ddot{r} + \frac{m}{r^2} - \frac{J^2}{r^3} + 3\frac{mJ^2}{r^4} = 0.$$

For a circular orbit at radius $r = R$, show that

$$J^2 = \frac{mR^2}{R - 3m}, \qquad \frac{d\varphi}{dt} = \left(\frac{m}{R^3}\right)^{1/2}.$$

Show by setting $r(\tau) = R + \epsilon(\tau)$, with ϵ small, that the circular orbit is stable if and only if $R > 6m$.

8.3. Show that for a suitable value of $\alpha = mE/J$, there are equatorial null geodesics in the Schwarzschild solution on which

$$\frac{1 - 3u}{\left(\sqrt{3} + \sqrt{1 + 6u}\right)^2} = Ae^\varphi$$

for arbitrary constant A. Describe their behaviour as $\varphi \to -\infty$ for (i) $A > 0$ and (ii) $A < 0$.

8.4. Sketch the phase portrait in the p, u-plane of the equatorial particle orbits in the Schwarzschild space–time for fixed E and various values of $\beta^2 = m^2/J^2$ in the case $1 > E^2 > 8/9$. What changes when $E^2 = 8/9$?

Black Holes

We now look more closely at what happens at the Schwarzschild radius, $r = 2m$. It is clear that something goes wrong there in the formula (7.6) for the metric coefficients. We show, however, that the singularity is not in the space–time geometry itself, but simply in the coordinates in which it is expressed. The singular behaviour at $r = 2m$ goes away when we make an appropriate change of coordinates.

9.1 The Schwarzschild Radius

For a normal star, the Schwarzschild radius is well inside the star itself. As it is not in the vacuum region of space–time, the Ricci tensor does not vanish at $r = 2m$, and so the Schwarzschild solution is not valid there. Instead the metric is that of an 'interior' Schwarzschild solution, found by solving Einstein's equations for a static spherically symmetric metric, with the energy-momentum tensor of an appropriate form of matter on the right-hand side. In such metrics, generally nothing exceptional happens at the Schwarzschild radius. But in the extreme case, all of the body lies within its Schwarzschild radius and the vacuum solution (7.6) extends down to $r = 2m$. In this case, we have a spherical *black hole*.

For the sun to be contained within its Schwarzschild radius, it would have to be compressed to a radius of 3 km, which would imply an almost unimaginable density. For a galaxy, however, the density at this critical compression is only that of air, and so it is not hard, at least in principle, to imagine a sufficiently advanced civilization directing the orbits of the stars in a galaxy so that all the matter ended up within the Schwarzschild radius. We must therefore take seriously the existence of black holes as a theoretical possibility even without having to contemplate the extreme conditions in which a star could collapse to a black hole.

9.2 Eddington–Finkelstein Coordinates

The Schwarzschild metric is

$$\mathrm{d}s^2 = \left(1 - \frac{2m}{r}\right)\mathrm{d}t^2 - \frac{\mathrm{d}r^2}{1 - 2m/r} - r^2(\mathrm{d}\theta^2 + \sin^2\theta\,\mathrm{d}\varphi^2).$$

We cannot simply ignore the part of space–time for which $r \leq 2m$ because an infalling observer will reach $r = 2m$ in finite proper time. An observer who falls radially, that is, with constant θ and φ, has worldline given by

$$E = (1 - 2m/r)\dot{t}, \qquad 1 = (1 - 2m/r)\dot{t}^2 - \frac{\dot{r}^2}{1 - 2m/r},$$

where the parameter τ is proper time. In the special case $E = 1$, which arises when the observer falls from rest with respect to the timelike Killing vector at infinity, we have $\dot{r}^2 = 2m/r$. Then

$$\int \sqrt{r}\,\mathrm{d}r = -\sqrt{2m} \int \mathrm{d}\tau$$

and hence

$$2r^{3/2} = 3\sqrt{2m}(\kappa - \tau)$$

for some constant κ. We conclude that the proper time τ taken to reach $r = 2m$ is finite. However, the coordinate time taken is infinite because

$$\frac{\mathrm{d}r}{\mathrm{d}t} = -\left(1 - \frac{2m}{r}\right)\frac{\sqrt{2m}}{\sqrt{r}}$$

and so

$$-\int \frac{r^{3/2}\,\mathrm{d}r}{r - 2m} = \sqrt{2m} \int \mathrm{d}t.$$

The integral on the left-hand side diverges as $r \to 2m$.

To understand the space–time geometry of a black hole, we first look for a coordinate system in which the singularity at $r = 2m$ disappears. One can see what goes wrong with the given coordinates by looking at the null geodesics in the r, t-plane—the worldlines of photons travelling radially inwards or outwards. These are the curves given by

$$\left(1 - \frac{2m}{r}\right) \mathrm{d}t^2 - \frac{\mathrm{d}r^2}{1 - 2m/r} = 0\,.$$

By integration we obtain

$$\int \mathrm{d}t = \pm \int \frac{\mathrm{d}r}{1 - 2m/r} = \pm \int \left(1 + \frac{2m}{r - 2m}\right) \mathrm{d}r\,.$$

That is,

$$t \pm \big(r + 2m\log(r - 2m)\big) = \text{constant}\,. \tag{9.1}$$

The radial null geodesics in the t, r-plane are the curves shown in Figure 9.1.

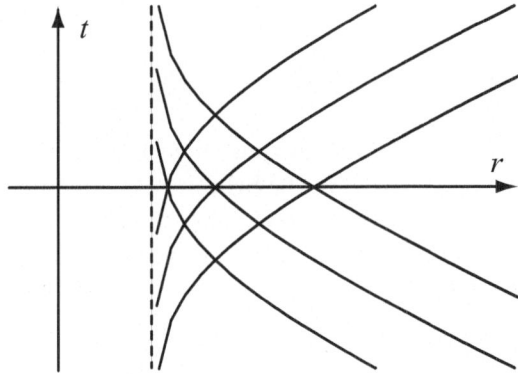

Figure 9.1 Radial null geodesics in the Schwarzschild metric

They all have $r = 2m$ as an asymptote, shown as a dashed line, and the singular behaviour there is associated with the fact that the curves bunch up on this common value of r. For large r, they look like the corresponding lines $r = \pm t$ in flat space–time. Each curve in Figure 9.1 represents an ingoing or outgoing spherical wavefront. One can get at least a partial picture of how this works by rotating about the t-axis to make the curves into surfaces of revolution. They are shown in Figure 9.1, which is a space–time diagram with one spatial dimension suppressed. The dark cylindrical surface is at $r = 2m$. The histories of outgoing and ingoing wavefronts are surfaces asymptotic to this.

We should compare this picture with the corresponding one for Minkowski space, where the radial null geodesics are straight diagonal lines in the t, r-plane at 45° and the corresponding in- and outgoing wavefronts are the null cones of the points on the polar axis $r = 0$ (Figure 9.3). In Minkowski space, as we

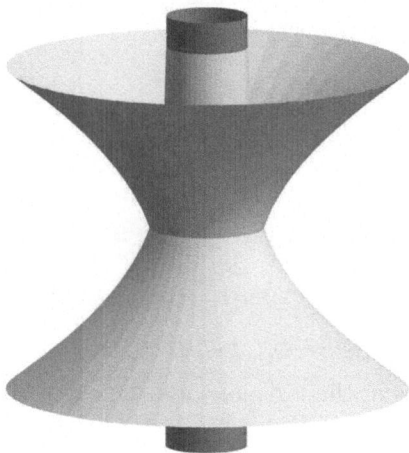

Figure 9.2 Ingoing and outgoing wavefronts in the Schwarzschild metric

follow an outgoing wavefront back in time, it focuses at the vertex of a cone, with the vertex lying on the axis. In the Schwarzschild picture, by contrast, the outgoing wavefront becomes closer and closer to the horizon as we follow it back into the past, without ever crossing it. The picture for the ingoing wavefronts is similar, but with time reversed.

Figure 9.3 Ingoing and outgoing wavefronts in Minkowski space

In the Schwarzschild space–time, we can resolve the coordinate difficulties by 'compressing' the t coordinate as we approach $r = 2m$. Guided by (9.1), we make the transformation to coordinates v, r, θ, φ by putting

$$v = t + r + 2m \log(r - 2m) \,,$$

which gives

$$dt = dv - \frac{dr}{1 - 2m/r}$$

and hence

$$ds^2 = (1 - 2m/r)\, dv^2 - 2dv\, dr - r^2(d\theta^2 + \sin^2\theta\, d\varphi^2) \,.$$

The singular behaviour at $r = 2m$ has now disappeared. In the r, v-plane, the

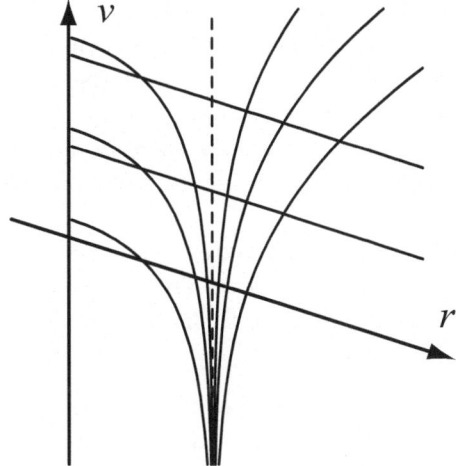

Figure 9.4 Radial null geodesics in Eddington–Finkelstein coordinates

radial null geodesics are the lines of constant v together with the solutions to

$$\left(1 - \frac{2m}{r}\right) \frac{dv}{dr} - 2 = 0 \,.$$

This can be integrated to give

$$v = \int \frac{2r\, dr}{r - 2m} = 2r + 4m \, \log |r - 2m| + \kappa \,, \tag{9.2}$$

for some constant κ. Thus the radial null geodesics are as shown in Figure 9.4. Because the r-axis itself is null, it is not drawn horizontally: the lines parallel to it are the lines of constant v. The null geodesics given by (9.2) have a common asymptote in the dashed vertical line.

We can see from the fact that timelike curves must lie between the ingoing and outgoing null geodesics at every event that although the space–time is nonsingular for $r < 2m$, it is not possible to escape to infinity. The hypersurface $r = 2m$ is called the *event horizon*. It separates events of which observers outside can have knowledge from those inside of which they cannot. The events inside the event horizon are inside the 'black hole'. The lines of constant v are null.

The new coordinates are called *Eddington–Finkelstein coordinates*. The histories of the ingoing and outgoing wavefronts outside the event horizon in Eddington–Finkelstein coordinates are shown in the space–time diagram, Figure 9.5. The dark cylinder is the horizon; the ingoing wavefront crosses the horizon, and the outgoing one is asymptotic to it in the past.

The singular behaviour of the metric coefficients in the t, r coordinates does not arise from a singularity of the space–time geometry because it disappears in the v, r coordinates. Instead it arises from the singular behaviour of the transformation from v, r to t, r coordinates at $r = 2m$, which shows itself in the fact that the curves of constant t are asymptotic to the line $r = 2m$ in the r, v-plane. The transformation from v, r to t, r coordinates pushes the points $(v, 2m)$ to $t = \infty$.

In Eddington–Finkelstein coordinates, the space–time extends to $r < 2m$. The Killing vector T with components $(1, 0, 0, 0)$ in the original coordinates t, r, θ, φ has the same components in the new coordinates, but inside the event horizon, it is spacelike. We have $T^a T_a = 1 - 2m/r$ and hence the following.

(i) For $r > 2m$, T is timelike and defines a standard of 'rest'. A stationary observer is one whose four-velocity is tangent to T.

(ii) For $r = 2m$, T is null. We can think of the event horizon as the history of a light wavefront 'at rest', hovering forever between escaping to infinity and falling into the black hole.

(iii) For $r < 2m$, T is spacelike, and no observer can remain at rest.

The worldline of any observer inside the black hole must inevitably reach $r = 0$ in finite proper time, in fact, in a time of the same order of magnitude as light takes to travel the Schwarzschild radius.

We cannot, however, extend beyond $r = 0$, whatever coordinates are used. There is a genuine singularity at $r = 0$, at which the tidal forces become infinite. One can see this from the fact the invariant $R_{abcd}R^{abcd}$ blows up like r^{-6}, and so there is no coordinate system in which the metric is well-behaved at $r = 0$. Once inside the black hole, an observer is not only unable to escape to infinity,

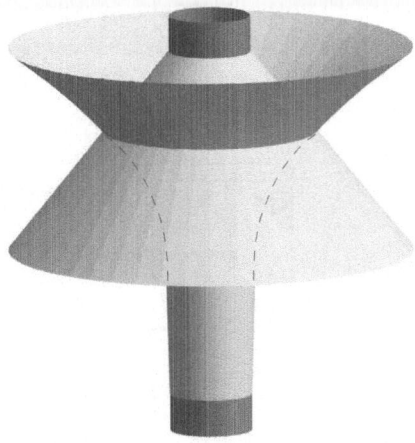

Figure 9.5 Wavefronts in Eddington–Finkelstein coordinates

but is also unable to escape being crushed in the singularity in a very short time.

9.3 Gravitational Collapse

The Schwarzschild solution by itself does not provide a good model of a real black hole because it is a vacuum metric. There is no matter present to generate the gravitational field. In a real astrophysical situation one expects black holes to form from the collapse of stars after they have burnt up all their nuclear fuel. The collapse can form a white dwarf, which is supported against gravity by the 'electron degeneracy pressure'; however, above 1.4 times the mass of the sun, this pressure is insufficient, and collapse results in a neutron star, essentially a massive nucleus with an atomic number around 10^{58}. But again there is a limit to mass. Above some critical mass, somewhere between 1.5 and 3 solar masses, no known physical process can prevent collapse to a black hole; and once the event horizon has formed, no conceivable process can prevent collapse to a singularity. This is the Penrose singularity theorem.

One can model the field of a spherically symmetric collapsing object by joining the Schwarzschild metric, to represent the field outside the body, to an interior metric, representing the field inside the collapsing star, across a spherically symmetric hypersurface represented by a timelike curve in the v, r-

plane. If we include one of the other spatial coordinates by rotating about the line $r = 0$, then we obtain the three-dimensional representation of the space–time shown in Figure 9.6.

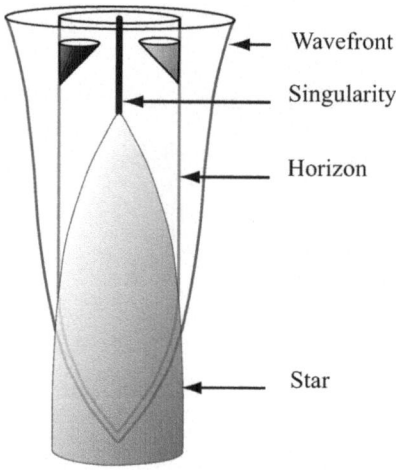

Figure 9.6 The collapse of a star to form a black hole

9.4 Kruskal Coordinates

It is instructive to explore further the vacuum solution without joining on any interior solution. Here we look more closely at a curious feature of the Eddington–Finkelstein coordinates, that they introduce a time asymmetry that is not present in the original metric. That is, they do not treat the future and the past in an even-handed way. They adjoin the interior of a black hole to the exterior solution. We could equally well reverse t and use the coordinate transformation to adjoin a 'white hole', from which an observer can escape, but cannot enter.

We can see what is going on here by transforming instead to *Kruskal coordinates*, in which both extensions can be made simultaneously. We start with the original form of the metric

$$ds^2 = \left(1 - \frac{2m}{r}\right) dt^2 - \frac{dr^2}{1 - 2m/r} - r^2 (d\theta^2 + \sin^2\theta \, d\varphi^2).$$

But now we transform to new coordinates U, V, θ, φ by putting

$$\frac{V}{U} = -\mathrm{e}^{t/2m}, \qquad UV = \mathrm{e}^{r/2m}(2m - r).$$

That is, $V = \mathrm{e}^{v/4m}$, $U = -\mathrm{e}^{-u/4m}$, where

$$v = t + r + 2m \log(r - 2m), \qquad u = t - r - 2m \log(r - 2m).$$

Here v is the Eddington–Finkelstein coordinate, and $-u$ is the coordinate used in the time-reversed extension. We then have

$$\mathrm{d}V = \frac{\mathrm{e}^{v/4m}}{4m}\left(\mathrm{d}t + \frac{\mathrm{d}r}{1 - 2m/r}\right), \qquad \mathrm{d}U = \frac{\mathrm{e}^{-u/4m}}{4m}\left(\mathrm{d}t - \frac{\mathrm{d}r}{1 - 2m/r}\right).$$

Hence

$$\mathrm{d}U\,\mathrm{d}V = \frac{r\,\mathrm{e}^{r/2m}}{16m^2}\left(1 - \frac{2m}{r}\right)\left(\mathrm{d}t^2 - \frac{\mathrm{d}r^2}{(1 - 2m/r)^2}\right).$$

Therefore in these new coordinates, the metric is

$$\mathrm{d}s^2 = 16m^2 r^{-1}\mathrm{e}^{-r/2m}\,\mathrm{d}U\mathrm{d}V - r^2(\mathrm{d}\theta^2 + \sin^2\theta\,\mathrm{d}\varphi^2),$$

where r is defined as a function of U, V by $UV = \mathrm{e}^{r/2m}(2m - r)$.

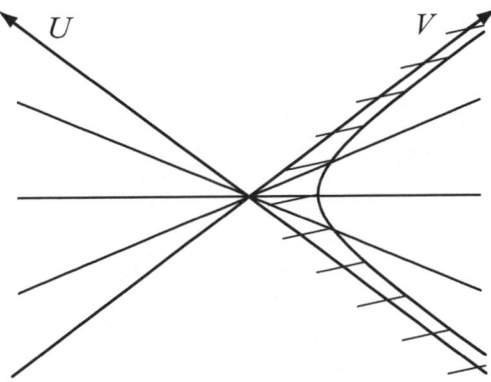

Figure 9.7 The U, V coordinates on Minkowski space

To understand the geometry, let us look first at the corresponding transformation of Minkowski space. Here we start with

$$\mathrm{d}s^2 = \mathrm{d}t^2 - \mathrm{d}r^2 - r^2(\mathrm{d}\theta^2 + \sin^2\theta\,\mathrm{d}\varphi^2)$$

and make the coordinate change

$$U = -\mathrm{e}^{r-t}, \qquad V = \mathrm{e}^{t+r}.$$

Then the metric becomes

$$\begin{aligned} ds^2 &= e^{-2r}\, dU\, dV - r^2(d\theta^2 + \sin^2\theta\, d\varphi^2) \\ &= -\frac{dU\, dV}{UV} - r^2(d\theta^2 + \sin^2\theta\, d\varphi^2)\,. \end{aligned}$$

If we suppress the angular coordinates, then the relationship between the two coordinate systems is as shown in Figure 9.7. The U, V axes are null lines, and are therefore drawn at 45° to the horizontal, with time and the two coordinates U, V increasing up the page. The curves of constant t are straight lines through the origin; those of constant r are the hyperbolas $UV =$ constant, which have the U, V-axes as asymptotes. The transformation maps the whole of Minkowski space into the region

$$-UV > 1, \qquad U < 0, \qquad V > 0$$

in the U, V-plane. The hyperbola in Figure 9.7 is the curve $UV = -1$; that is, $r = 0$. The excluded region is the shaded region to the left of the right-hand branch.

In the Schwarzschild geometry, the picture is very similar, except that the metric continues in the U, V-plane to the region $UV < 2m$. The boundary $UV = 2m$ is the image of the 'real' singularity at $r = 0$ in the r, t plane. Figure 9.8 again shows the U, V-plane, with the axes drawn at 45° to the horizontal. The straight lines are null. In this case, however, the metric is nonsingular in

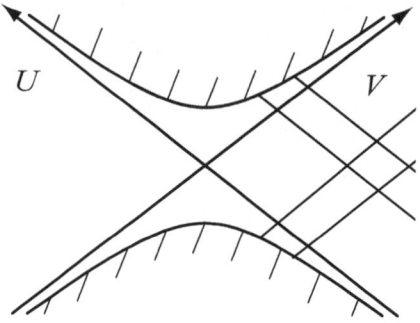

Figure 9.8 The Kruskal extension of the Schwarzschild geometry

the whole region bounded by the two branches of the hyperbola $UV = 1$, on which $r = 0$. If we exclude the shaded region above and below these, then we have the *maximal analytic extension* of the Schwarzschild space–time. The portion covered by the Eddington–Finkelstein coordinates is the portion above the U-axis. The entire extended space–time contains both a black hole, the region $U > 0$, $V > 0$, which an observer can enter but not leave, and a 'white hole'—the time reverse of a black hole—the region $U < 0$ $V < 0$, which an observer can leave but not enter. There is no matter present. We can think of the 'm' in the metric as being entirely gravitational in origin, or perhaps we should think of it as the mass of the singularity at $r = 0$. There is no stellar boundary and the space–time looks like two external regions, joined by a 'wormhole'.

The external regions are the two quadrants $V > 0 > U$ and $U > 0 > V$: for large $|UV|$, the metric looks in both like that of Minkowski space. We can see the way in which they are connected by looking at the geometry of the spatial section $U = V$, on which r is given as a function of V by $V^2 = \mathrm{e}^{r/2m}(2m - r)$. On this r decreases to a minimum value of $2m$ and then increases again to

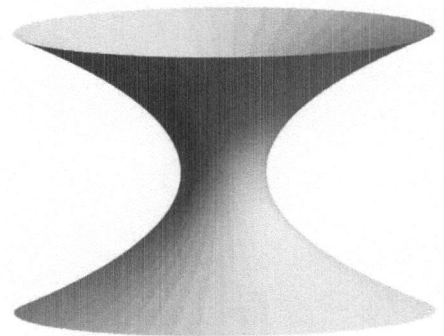

Figure 9.9 The spatial geometry at $t = 0$

infinity. If we put $\theta = \pi/2$ (so that we are looking at the 'equatorial plane'), then the metric is

$$\mathrm{d}s^2 = (1 - 2m/r)^{-1}\,\mathrm{d}r^2 + r^2\,\mathrm{d}\varphi^2 = (1 + f'(r)^2)\mathrm{d}r^2 + r^2\mathrm{d}\varphi^2\,,$$

where $f = \sqrt{8m(r - 2m)}$. This is the metric on a surface of revolution given by rotating the parabola $f = f(r)$ about the f-axis. Thus we can picture the hypersurface $U = V$ as two copies of Euclidean space (at large r), joined by the tube in Figure 9.9. This is the *wormhole*. To an observer in either of the external spaces, the geometry looks like that of a black hole. Of course one cannot actually travel through the wormhole. The passage through $r = 2m$ takes one inside the event horizon, and inevitably into the singularity at $r = 0$.

When the black hole is formed by gravitational collapse, we see only part of
the diagram to the right in Figure 9.8. The rest must be replaced by a suitable
interior metric.

10

Rotating Bodies

The Schwarzschild metric gives us some of the classic tests of relativity: the bending of light, Mercury's perihelion precession, and other predictions from the analysis of geodesic motion. It also allows us to make some dramatic predictions about the end states of the gravitational collapse of stars to black holes. To find deeper tests, we have to look for more subtle effects of general relativity, which cannot be seen in the Schwarzschild space–time. One is the 'dragging of inertial frames' by a rotating body. The predictions here allow the testing of Einstein's equations as well as of the geometric model of space–time. They can be observed in the effect of the earth's rotation on an orbiting gyroscope.

We find the weak-field metric outside a rotating body before considering the frame-dragging effect. We then look briefly at the Kerr metric, which is an exact solution for the field.

10.1 The Weak Field Approximation

We begin with Einstein's equations in the form

$$R_{ab} - \tfrac{1}{2}Rg_{ab} = -8\pi T_{ab},$$

where $T_{ab} = \rho U_a U_b$ is the energy-momentum tensor of a distribution of dust with rest density ρ and four-velocity field U^a. In the weak field approximation (§6.2), $g_{ab} = m_{ab} + h_{ab}$ and

$$R_{abc}{}^d = \tfrac{1}{2}m^{de}(\partial_a\partial_c h_{be} + \partial_b\partial_e h_{ac} - \partial_a\partial_e h_{bc} - \partial_b\partial_c h_{ae}).$$

Therefore to the same approximation

$$R_{ab} = \tfrac{1}{2}\left(\Box h_{ab} - \partial_a Y_b - \partial_b Y_a\right).$$

where $Y_c = m^{ab}(\partial_a h_{bc} - \tfrac{1}{2}\partial_c h_{ab})$ and $\Box$ is the d'Alembertian. Here m_{ab} is the metric on a background Minkowski space and the coordinates x^a are inertial.

There is one obvious coordinate freedom in this 'linearized' form of Einstein's theory, which is to make a Lorentz transformation. There is also a less obvious one, which can be seen as a *gauge transformation* in the weak field theory. The idea is to replace the x^as by

$$x^a + Z^a,$$

where Z^a is a vector field with small components, of the same order as those of h_{ab}. The effect is to transform m_{ab} to

$$m_{ab} + \partial_a Z_b + \partial_b Z_a,$$

where the $Z_a = m_{ab}Z^b$. In the spirit of the original approximation, we have dropped terms involving products of the derivatives of Z^a. We can absorb the change in m_{ab} into h_{ab} by making the *gauge transformation*

$$h_{ab} \mapsto h_{ab} + 2\partial_{(a} Z_{b)}. \tag{10.1}$$

We then again have a 'weak field' deviation from flat space–time. So part of the perturbation of m_{ab} can be seen as a perturbation in the background inertial coordinates and part as a genuine gravitational field. In general, there is no natural way to disentangle the two.

We can, however, exploit the gauge freedom to restrict the form of h_{ab}. In particular we can impose the *de Donder gauge condition* $Y_a = 0$. Under (10.1),

$$Y_c = m^{ab}(\partial_a h_{bc} - \tfrac{1}{2}\partial_c h_{ab}) \mapsto Y_c + \Box Z_c.$$

So to find a transformation that makes h_{ab} vanish, it is necessary only to choose the Z_as to be solutions of the inhomogeneous wave equation $\Box Z_a = -Y_a$.

In the de Donder gauge, the approximate form of Einstein's equation is

$$\Box w_{ab} = -16\pi T_{ab}, \tag{10.2}$$

where

$$w_{ab} = h_{ab} - \tfrac{1}{2}m_{ab}m^{cd}h_{cd}.$$

Exercise 10.1

Show that the approximate curvature (6.4) is invariant under gauge transformations. Thus the observable effects of the gravitational field are unaltered.

Example 10.1 (Linearized Schwarzschild metric)

Identify the coordinates in the Schwarzschild metric (7.6) with spherical polar coordinates in Minkowski space. If m is small and if we ignore terms of order m^2, then the metric reduces in inertial coordinates to

$$ds^2 = dt^2 - dx^2 - dy^2 - dz^2 - 2mr^{-1}(dt^2 + dr^2),$$

with r defined by $r^2 = x^2 + y^2 + z^2$. The metric perturbation $-2mr^{-1}(dt^2 + dr^2)$ is not in de Donder gauge. But a gauge transformation by

$$(Z_a) = -mr^{-1}(0, x, y, z)$$

puts it in this gauge, with

$$(h_{ab}) = -\frac{2m}{r} \begin{pmatrix} 1 & 0 & 0 & 0 \\ 0 & 1 & 0 & 0 \\ 0 & 0 & 1 & 0 \\ 0 & 0 & 0 & 1 \end{pmatrix}, \qquad (w_{ab}) = -\frac{4m}{r} \begin{pmatrix} 1 & 0 & 0 & 0 \\ 0 & 0 & 0 & 0 \\ 0 & 0 & 0 & 0 \\ 0 & 0 & 0 & 0 \end{pmatrix}.$$

10.2 The Field of a Rotating Body

Suppose that the gravitational field is time-independent and is generated by a distribution of slow-moving matter with small density ρ and velocity field $\boldsymbol{u}$, with $u \ll 1$. Then $T^{ab} = \rho U^a U^b$, with $(U^a) \sim (1, u_1, u_2, u_3)$. Equation (10.2) takes the form

$$\nabla^2 w_{00} = 16\pi\rho, \qquad \nabla^2 w_{0i} = -16\pi\rho u_i, \qquad \nabla^2 w_{ij} = 16\pi\rho u_i u_j, \qquad (10.3)$$

where $i, j = 1, 2, 3$ and ∇^2 is the Laplacian of the spatial coordinates. The right-hand side of the third equation is quadratic in small quantities, and thus is ignored. Therefore we can put

$$w_{ij} = 0 \qquad i, j = 1, 2, 3$$

in this approximation. The first equation gives

$$w_{00}(\boldsymbol{r}) = -4 \int \frac{\rho(\boldsymbol{r}') \, dV'}{|\boldsymbol{r} - \boldsymbol{r}'|},$$

where the integral is over the matter, $\boldsymbol{r}' = (x', y', z')$ is the position vector of a volume element dV', and $\boldsymbol{r} = (x, y, z)$ is the point at which the metric component is evaluated. If we take the origin at the centre of mass and assume that the size of the body is small compared with distance r from the centre, then this gives

$$w_{00} = -\frac{4m}{r} + O(r^{-2}).$$

Because $m^{ab}h_{ab} = -m^{ab}w_{ab} = -w_{00}$, we then get

$$h_{00} = h_{11} = h_{22} = h_{33} = 2\phi + O(r^{-2}),$$

where $\phi = -m/r$ is the Newtonian potential, together with $h_{ij} = 0$ when $i \neq j$.

To find the remaining components of h_{ab}, we make the further simplifying assumption that the body is a rigid sphere rotating with angular velocity $\boldsymbol{\omega}$ and with a spherically symmetric distribution of matter. This is not consistent with the dust form of the energy-momentum tensor, but the gravitational effect of the internal stresses is negligible. With this assumption, the velocity of the point with position vector $\boldsymbol{r}'$ is $\boldsymbol{u} = \boldsymbol{\omega} \wedge \boldsymbol{r}'$. Consider the component $h_{01} = w_{01}$. From (10.3), this is

$$w_{01}(\boldsymbol{r}) = 4 \int \frac{\rho(\boldsymbol{r}')u_1(\boldsymbol{r}')\,\mathrm{d}V'}{|\boldsymbol{r} - \boldsymbol{r}'|} = 4 \int \frac{\rho(\boldsymbol{r}')(\omega_2 z' - \omega_3 y')\,\mathrm{d}V'}{|\boldsymbol{r} - \boldsymbol{r}'|};$$

but we have

$$\frac{1}{\sqrt{(\boldsymbol{r} - \boldsymbol{r}') \cdot (\boldsymbol{r} - \boldsymbol{r}')}} = \frac{1}{r} + \frac{xx' + yy' + zz'}{r^3} + O(r^{-3}).$$

Because the origin is at the centre of the sphere, the integrals of

$$\rho(\boldsymbol{r}')x', \quad \rho(\boldsymbol{r}')y', \quad \rho(\boldsymbol{r}')z', \quad \rho(\boldsymbol{r}')y'z', \quad \rho(\boldsymbol{r}')z'x', \quad \rho(\boldsymbol{r}')x'y'$$

over the sphere all vanish, and

$$\int \rho(\boldsymbol{r}')x'^2\,\mathrm{d}V' = \int \rho(\boldsymbol{r}')y'^2\,\mathrm{d}V' = \int \rho(\boldsymbol{r}')z'^2\,\mathrm{d}V' = \tfrac{1}{2}I,$$

where I is the moment of inertia of the sphere about its centre. Hence

$$\begin{aligned} h_{01} &= 4 \int \frac{\rho(\boldsymbol{r}')(\omega_2 z' - \omega_3 y')(xx' + yy' + zz')\,\mathrm{d}V'}{r^3} + O(r^{-3}) \\ &= 2r^{-3}I(\boldsymbol{\omega} \wedge \boldsymbol{r})_1 + O(r^{-3}). \end{aligned}$$

From this and the similar calculation for the other two components, we conclude that $h_{0i} = \alpha_i$, with $\boldsymbol{\alpha}$ defined by

$$\boldsymbol{\alpha} = 2\boldsymbol{L} \wedge \boldsymbol{r}/r^3, \tag{10.4}$$

where $\boldsymbol{L}$ is the angular momentum about the centre of mass. To within our approximation, therefore, the metric outside the rotating body is

$$\mathrm{d}s^2 = (1 + 2\phi)\,\mathrm{d}t^2 + 2\mathrm{d}t\,\boldsymbol{\alpha} \cdot \mathrm{d}\boldsymbol{r} - (1 - 2\phi)\,\mathrm{d}\boldsymbol{r} \cdot \mathrm{d}\boldsymbol{r}, \tag{10.5}$$

where $d\boldsymbol{r} = (dx, dy, dz)$, $\phi = -M/r$ is the Newtonian gravitational potential, $\boldsymbol{\alpha}$ is related to the angular momentum $\boldsymbol{L}$ of the gravitating source by (10.4), and the dot is the usual dot product in Euclidean space.

Exercise 10.2

Let T be the timelike Killing vector in (10.5). Find $\nabla_{[a} T_{b]} \neq 0$ in terms of $\boldsymbol{\alpha}$.

10.3 The Lens–Thirring Effect

The effect of the angular momentum term in (10.5) can be seen in the precession, or rotation of the axis, of a gyroscope carried in free-fall. The effect, known as the *Lens–Thirring effect*, is often interpreted as being the result of the *dragging of local inertial frames* by the rotating body. As always in relativity it is necessary to be clear about the precise meaning of statements involving motion and rotation. Neither the prediction of precession nor the interpretation in terms of dragging make sense without spelling out what is rotating relative to what.

A gyroscope is an axisymmetric body rotating about its axis of symmetry. The Newtonian angular momentum conservation law implies that, in the absence of forces, the direction of the axis is constant. What happens in a gravitational field? Suppose that the gyroscope is carried by an observer in free-fall. Our central principle that classical theory should hold good in free-fall over short times and distances implies that the direction of the axis should remain constant relative to local inertial coordinates.

We can put this statement in a more convenient form. Denote the observer's four-velocity by V and, at each event on the observer's worldline, let E denote the spacelike vector with components $(0, e_1, e_2, e_3)$ in local inertial coordinates at the event, where $\boldsymbol{e}$ is the unit vector along the axis of the gyroscope. Then

$$E_a E^a = -1, \qquad E_a V^a = 0, \tag{10.6}$$

and the statement that the direction of the axis is constant in local inertial coordinates translates to

$$DE^a = 0,$$

where D is the covariant derivative along the worldline. Note that $E_a V^a$ is constant because $DV^a = 0$ as a result of the geodesic equation.

It is easy to understand in physical terms what is meant by precession if one imagines the observer being in orbit about the earth and comparing the direction of the axis of the gyroscope with the directions to fixed stars. The statement is that e is seen to rotate relative to stars, the rotation being made up of one element—'geodetic precession'—that can be found from the Newtonian potential, and a rather smaller one—the Lens–Thirring term—which involves the angular momentum of the earth. In mathematical terms, we need to understand 'change in direction relative to the fixed stars' in terms of a procedure for comparing the values of E^a at different events on the worldline.

The key is the timelike Killing vector T^a of the weak field metric (10.5). If we have a second observer at rest relative to T, and if the first passes the second at two events A and B with the same relative speed, then we can compare the values of E at A and B in an unambiguous way by comparing its components at the two events in the coordinates of (10.5). The following enables us to calculate the change.

We assume that ϕ and the free-falling observer's velocity relative to a stationary observer are small. We keep quadratic terms in these small quantities and their derivatives but ignore cubic and smaller terms. We also assume that the metric perturbation components h_{0i} are very much smaller than h_{00}, and therefore we also ignore terms involving the product of $\boldsymbol{\alpha}$ with ϕ or with the relative speed.

We write a four-vector X in the coordinate system of (10.5) as $(\xi, \boldsymbol{x})$, where $\xi = X^0$ and $\boldsymbol{x} = (X^1, X^2, X^3)$. We then have

$$X^a X_a = (1 + 2\phi)\xi^2 - (1 - 2\phi)\boldsymbol{x} \cdot \boldsymbol{x} + 2\xi\,\boldsymbol{\alpha} \cdot \boldsymbol{x},$$

where the dot is the standard inner product, defined by $\boldsymbol{a} \cdot \boldsymbol{b} = a_1 b_1 + a_2 b_2 + a_3 b_3$. In this notation, the four-velocity V of the free-falling observer is a scalar multiple of

$$W = (1, \boldsymbol{v}) = \frac{\mathrm{d}}{\mathrm{d}t}(t, \boldsymbol{r}),$$

where $\boldsymbol{v} = \mathrm{d}\boldsymbol{r}/\mathrm{d}t$. So we can deduce from (10.6) that

$$E = \left(\boldsymbol{z} . \boldsymbol{v}, \boldsymbol{z} + \phi \boldsymbol{z} + \tfrac{1}{2}(\boldsymbol{z} \cdot \boldsymbol{v})\,\boldsymbol{v}\right)$$

for some $\boldsymbol{z}$ such that $\boldsymbol{z} \cdot \boldsymbol{z} = 1$. In computing the inner products $E^a W_a$ and $E^a E_a$, we keep only the terms of the same order as ϕ or v^2.

We want to find $\mathrm{d}\boldsymbol{z}/\mathrm{d}t$ as E is parallel transported along the worldline. We do this by writing the equation of parallel transport in the form

$$\frac{\mathrm{d}E^a}{\mathrm{d}t} + \Gamma^a_{bc} W^b E^c = 0 \tag{10.7}$$

in the coordinates of (10.5); W appears here rather than V because the parameter is the coordinate time t. The ith term in (10.7) is

$$\frac{\mathrm{d}}{\mathrm{d}t}\left((1 + \phi)z_i + \tfrac{1}{2}z_j v_j v_i\right) + \Gamma^i_{00} v_j z_j + \Gamma^i_{0j} z_j + \Gamma^i_{jk} v_j z_k = 0, \tag{10.8}$$

with summation over $j, k = 1, 2, 3$. Now

$$\frac{\mathrm{d}\phi}{\mathrm{d}t} = \boldsymbol{v} \cdot \boldsymbol{\nabla}\phi \quad \text{and} \quad \frac{\mathrm{d}\boldsymbol{v}}{\mathrm{d}t} = -\boldsymbol{\nabla}\phi,$$

as in Newtonian theory. From (6.3), we have for $i, j, k = 1, 2, 3$,

$$\Gamma^i_{00} = \partial_j\phi, \qquad \Gamma^i_{0j} = \tfrac{1}{2}(\partial_i\alpha_j - \partial_j\alpha_i), \qquad \Gamma^i_{jk} = \partial_i\phi\,\delta_{jk} - \partial_j\phi\,\delta_{ik} - \partial_k\phi\delta_{ij}\,.$$

Therefore (10.8) is the ith component of

$$\frac{\mathrm{d}\boldsymbol{z}}{\mathrm{d}t} - \tfrac{3}{2}(\boldsymbol{z}\cdot\boldsymbol{\nabla}\phi)\boldsymbol{v} + \tfrac{3}{2}(\boldsymbol{v}\cdot\boldsymbol{z})\boldsymbol{\nabla}\phi + \tfrac{1}{2}\boldsymbol{z}\wedge\operatorname{curl}\boldsymbol{\alpha} = 0\,.$$

It follows that

$$\frac{\mathrm{d}\boldsymbol{z}}{\mathrm{d}t} = \left(\tfrac{3}{2}\boldsymbol{\nabla}\phi\wedge\boldsymbol{v} + \tfrac{1}{2}\operatorname{curl}\boldsymbol{\alpha}\right)\wedge\boldsymbol{z}\,.$$

Thus $\boldsymbol{z}$ rotates with angular velocity $\boldsymbol{\omega} = \tfrac{3}{2}\boldsymbol{\nabla}\phi\wedge\boldsymbol{v} + \tfrac{1}{2}\operatorname{curl}\boldsymbol{\alpha}$.

How should we interpret this rotation? We want think of $\boldsymbol{z}$ as the vector in the background flat space–time that 'points in the same direction' as the axis of the gyroscope. The difficulty with this is that there is no natural way to separate the space–time geometry into a background flat space–time metric and a small perturbation h_{ab}, because of the coordinate gauge freedom. If we want to interpret the rotation, for example, as being relative to the 'fixed stars' then we have to take account of the fact that light from distant stars does not travel in straight lines in the x, y, z coordinates because of the bending of light.

We can, however, apply the calculation of $\boldsymbol{\omega}$ to find a rotation that has an unambiguous interpretation when the free-falling observer is on a closed orbit which returns periodically to the same position, measured by x, y, z, at the same velocity. This is the context in which the prediction is being put to the test. We take the free-fall worldline to be the history of a satellite in orbit around the earth, and we model the earth's gravitational field by the metric (10.5). After each complete orbit, the satellite returns to the same position and velocity. The relationship between E^a and $\boldsymbol{z}$ is the same at each return, so any rotation in $\boldsymbol{z}$ between each return is unambiguously a real effect of the gravitational field: it will be observed as a rotation of the axis of the gyroscope relative to the apparent position of stationary stars. It is true, of course, that the satellite does not return to exactly the same position and velocity in general relativity, as we saw in the derivation of the perihelion advance; but that effect is negligible in this context.

The rotation has two components: a larger one

$$\tfrac{3}{2}\boldsymbol{\nabla}\phi\wedge\boldsymbol{v}\,,$$

called the *geodetic precession*, which was predicted by de Sitter in 1916, shortly after the first publication of general relativity. It has been observed by treating the earth–moon system as a gyroscope in free-fall in the field of the sun [15]. The second,

$$\tfrac{1}{2}\operatorname{curl}\boldsymbol{\alpha}\,,$$

is the smaller *Lens–Thirring precession*, which is currently being measured directly by Gravity Probe B, by measuring the cumulative change in direction of the axis of a gyroscope in a circular polar orbit against the fixed stars over many orbits. This is more sensitive than the geodetic precession to the differences between Einstein's theory and other possible theories of gravity. The rates of precession in this context are 6.6 seconds of arc per year for the geodetic precession and 0.041 seconds of arc per year for the Lens–Thirring precession. Extraordinary ingenuity and precision are needed to separate the latter from the former. The paper by Lämmerzahl and Neugebauer [12] gives a detailed discussion of the history and theoretical background and a derivation of these rates of rotation. For an account of Gravity Probe B, see [6].

10.4 The Kerr Metric

The metric (10.5) models the approximate field outside a rotating body with angular momentum $\boldsymbol{L}$. In 1963 Kerr found an exact solution to this problem, in the form of the *Kerr metric* [11]. In *Boyer–Lindquist coordinates*, it is

$$\mathrm{d}t^2 - \frac{2mr}{\Sigma}\left(a\sin^2\theta\,\mathrm{d}\varphi - \mathrm{d}t\right)^2 - \Sigma\left(\mathrm{d}\theta^2 + \frac{\mathrm{d}r^2}{\Delta}\right) - (r^2 + a^2)\sin^2\theta\,\mathrm{d}\varphi^2\,, \quad (10.9)$$

where a, m are constant, and

$$\Delta = r^2 - 2mr + a^2, \qquad \Sigma = r^2 + a^2\cos^2\theta\,.$$

For small a and m, the Kerr metric reduces to the approximate solution. On replacing r by $r - m$ and on dropping terms in a^2 and m^2, (10.9) becomes

$$(1 - 2m/r)\mathrm{d}t^2 + 4mar^{-1}\sin^2\theta\,\mathrm{d}t\mathrm{d}\varphi - (1 + 2m/r)(\mathrm{d}r^2 - r^2\mathrm{d}\theta^2 - r^2\sin^2\theta\mathrm{d}\varphi^2)\,.$$

This is the same as the weak field metric (10.5) for a source with mass m and angular momentum am in the direction of the axis of the polar coordinates.

Exercise 10.3

Show that if r, θ, ϕ are spherical polar coordinates, then

$$2mar^{-1}\sin^2\theta\,\mathrm{d}\phi = \alpha_1\,\mathrm{d}x + \alpha_2\,\mathrm{d}y + \alpha_3\,\mathrm{d}z\,,$$

where $\boldsymbol{\alpha} = 2mar^{-3}(-y, x, 0)$. Hence by comparison with (10.5), show that in this weak field approximation, the Kerr metric has angular momentum am in the direction of the z-axis.

By analogy with the coordinate transformation in 9.2, we can replace the t and φ coordinates in (10.9) by v and ψ, where

$$v = t + \int \frac{(r^2 + a^2)\,\mathrm{d}r}{\Delta}, \qquad \psi = \varphi + \int \frac{a\,\mathrm{d}r}{\Delta}.$$

The metric then becomes

$$\begin{aligned}
\mathrm{d}s^2 &= \mathrm{d}v^2 - \frac{2mr}{\Sigma}(\mathrm{d}v - a\sin^2\theta\,\mathrm{d}\psi)^2 - 2\mathrm{d}v\,\mathrm{d}r - \Sigma\,\mathrm{d}\theta^2 \\
&\quad + 2a\sin^2\theta\,\mathrm{d}r\mathrm{d}\psi - (r^2 + a^2)\sin^2\theta\,\mathrm{d}\psi^2,
\end{aligned} \qquad (10.10)$$

without approximation.

The bold and energetic will calculate the Ricci tensor and show that it vanishes. It was not through this lengthy calculation that the solution was discovered, rather it was through seeking exact—not approximate—solutions of the *Kerr–Schild* form $g_{ab} = m_{ab} - n_a n_b$, where m_{ab} is the Minkowski metric and n_a is null. In fact, a further coordinate transformation

$$\begin{aligned}
\tilde{x} &= r\sin\theta\cos\psi - a\sin\theta\sin\psi, \\
\tilde{y} &= r\sin\theta\sin\psi + a\sin\theta\cos\psi, \\
\tilde{z} &= r\cos\theta, \\
\tilde{t} &= v - r
\end{aligned}$$

brings the Kerr metric into the form

$$g_{ab} = m_{ab} - \frac{2mr^3\,n_a n_b}{r^4 + a^2 z^2}$$

with

$$(n_0, n_1, n_2, n_3) = \left(1, \frac{r\tilde{x} - a\tilde{y}}{r^2 + a^2}, \frac{r\tilde{y} + a\tilde{x}}{r^2 + a^2}, \frac{\tilde{z}}{r}\right),$$

and r determined in terms of $\tilde{x}, \tilde{y}, \tilde{z}$ by the condition that n_a should be null with respect to the Minkowski metric. See [8].

Exercise 10.4

Show that n_a is also null with respect to the Kerr metric.

11
Gravitational Waves

In the last chapter, we saw that in the weak-field approximation, Einstein's equations for a perturbation of the Minkowski metric can be reduced to

$$\Box w_{ab} = -16\pi T_{ab} \,,$$

in the de Donder gauge [see (10.2)]. This is an inhomogeneous wave equation, with the energy-momentum tensor as source. It is strongly reminiscent of Maxwell's equations for the four-potential in the Lorenz gauge and it has the same implication. Maxwell's equations imply that moving charges generate electromagnetic waves. Einstein's equations imply that moving masses generate gravitational waves.

In this chapter we explore how this works, for the most part in the linearized theory. Gravitational waves have yet to be detected directly, although the predicted loss of energy through gravitational radiation in the binary pulsar PSR 1913+16 has been verified [21]. It is hoped that radiation from extreme astronomical events will be seen directly in the next few years by laser interferometry detectors [13]. The observations are very delicate because gravitational forces are many orders of magnitude weaker than electromagnetic ones. The electrostatic repulsion between two protons is a factor of 1.2×10^{36} greater than their gravitational attraction, at any separation: both forces obey the inverse square law. There are also formidable theoretical problems in understanding the generation of waves. We derive a form of Einstein's 'quadrupole formula' for wave production in the weak field theory. It is not at all straightforward, however, to take over this result into the full theory and to apply it in the astrophysical context in which it is needed. In the collision of two black holes, for example, the waves produced must escape from the vicinity of the black holes. The linearized theory does not tell us how they interact with the strong background field of the black holes themselves.

11.1 Metric Perturbations

In the linearized theory, one studies the behaviour of *metric perturbations* in a background Minkowski space. The space–time metric is $g = m_{ab} + h_{ab}$, where h_{ab} is a small perturbation of the background Minkowski metric m_{ab}, and

$$w_{ab} = h_{ab} - \tfrac{1}{2} m^{cd} h_{cd} m_{ab}\,.$$

That is, w_{ab} is the *trace reversal* of h_{ab}. The de Donder gauge condition is that

$$m^{ab} \partial_a w_{bc} = 0\,.$$

See §10.1. If we use the Minkowski metric m_{ab} and its inverse m^{ab} to lower and raise indices, then we can write more simply

$$w_{ab} = h_{ab} - \tfrac{1}{2} h m_{ab}, \qquad h = h_a{}^a, \qquad \partial_a w^{ab} = 0\,.$$

The gauge is fixed up to

$$h_{ab} \mapsto h_{ab} + \partial_a Z_b + \partial_b Z_a, \qquad \Box Z_a = 0\,.$$

This framework is closely analogous to the four-potential form of Maxwell's equations. In the Lorenz gauge, these are

$$\Box \Phi_a = k J_a, \qquad \partial_a \Phi^a = 0\,, \tag{11.1}$$

where Φ_a is the four-potential, J_a is the four-current, and k is a constant, equal to $1/c\epsilon_0$ in standard units. Maxwell's equations predict the existence of electromagnetic waves. Einstein's equations similarly predict the existence of gravitational waves.

11.2 Plane Harmonic Waves

In the absence of sources, we have

$$\Box w_{ab} = 0, \qquad \partial_a w^{ab} = 0\,, \tag{11.2}$$

so the individual components of w_{ab} satisfy the wave equation. These equations have *harmonic plane wave* solutions

$$w_{ab} = A_{ab} \cos(n_c x^c) + B_{ab} \sin(n_c x^c) , \qquad (11.3)$$

where n^a is a constant null vector and $A_{ab}n^b = B_{ab}n^b = 0$. We can write them more simply as

$$w_{ab} = \mathrm{Re}\left(k_{ab} \exp(-in_c x^c)\right) ,$$

where $k_{ab} = A_{ab} + iB_{ab}$ and Re denotes the real part. Under a gauge transformation with

$$Z_a = \mathrm{Re}\left(z_a \exp(-in_b x^b)\right) ,$$

where z_a is constant and complex, the observable properties of the linearized field are unchanged, but k_{ab} is replaced by

$$k_{ab} - 2in_{(a}z_{b)} + in^c z_c m_{ab} .$$

The complex tensor k_{ab}, subject to the condition $n^a k_{ab} = 0$, has six independent components. That number can be reduced to two by making a gauge transformation with an appropriate choice of z_a. In particular, one can always set $w = w_a{}^a = 0$, so that $w_{ab} = h_{ab}$, and both are traceless.

Exercise 11.1

Show that in addition it is always possible to choose the gauge of a harmonic plane wave so that $t^a h_{ab} = 0$, where t^a is the unit vector along the time axis of the inertial coordinates. This is the *transverse traceless* gauge.

Linear combinations of harmonic plane waves are the 'general solutions' of the linearized vacuum equation in the sense that any solution to (11.2) that falls off sufficiently quickly at infinity can be written in the form

$$w_{ab} = \mathrm{Re}\int k_{ab}(\boldsymbol{n}) \exp(-in_c x^c)\, dV' ,$$

where the integral is over all $\boldsymbol{n} \in \mathbb{R}^3$ and dV' is the volume element $dn_1\, dn_2\, dn_3$. The coefficient k_{ab} is a symmetric, complex-valued function of $\boldsymbol{n}$ and is orthogonal to n^a in the sense that $n^a k_{ab} = 0$; the real null vector n^a has spatial part $\boldsymbol{n}$ and temporal part $n^0 = \sqrt{\boldsymbol{n} \cdot \boldsymbol{n}}$.

The proof uses the inverse Fourier transform and the uniqueness theorem for the wave equation. With a dot denoting the partial derivative with respect to the inertial coordinate t, we have

$$k_{ab}(\boldsymbol{n}) = \frac{1}{(2\pi)^3} \int_{t=0} \frac{n^0 w_{ab} - i\dot{w}_{ab}}{n^0} \exp(-i\boldsymbol{n} \cdot \boldsymbol{r})\, dV$$

with integral over all $\boldsymbol{r} \in \mathbb{R}^3$ at $t = 0$ and $dV = dr_1,\, dr_2\, dr_3$.

Exercise 11.2

Prove this formula.

11.3 Plane and Plane-Fronted Waves

The harmonic plane wave (11.3) is a gravitational wave of a definite frequency travelling with the speed of light in the direction of the vector $\boldsymbol{n}$. The metric disturbance is of the form

$$h_{ab} = \mathrm{Re}\left(c_{ab}\exp(-\mathrm{i}n_d x^d)\right),$$

where $c_{ab} = k_{ab} - \frac{1}{2}k_d{}^d m_{ab}$ is constant, with complex components.

More generally a *plane wave* is a combination of harmonic waves all travelling in the same direction. It is a solution of (11.2) that depends on the inertial coordinates only through the combination $u = n_a x^a$ for some constant null four-vector n^a. The corresponding metric disturbance is characterized by the fact that

$$n^a w_{ab} = 0, \qquad X^a \partial_a w_{bc} = 0 \tag{11.4}$$

for every four-vector X^a such that $X^a n_a = 0$. Because these conditions are not preserved by gauge transformations, we also call a metric disturbance a 'plane wave' if it can be transformed to one satisfying these conditions by a change of gauge; that is, by the addition of $2\partial_{(a}Z_{b)}$ for some covector Z_a.

If w_{ab} satisfies the conditions (11.4), then $h_{ab} = w_{ab} - \frac{1}{2}w_c{}^c m_{ab}$ also depends only on u. An illuminating gauge transformation is given by putting

$$Z_a = \tfrac{1}{4}\left(h'_{bc}x^b x^c n_a - 2h_{ab}x^b\right),$$

where the prime is the derivative with respect to u. Then on replacing h_{ab} by $h_{ab} + 2\partial_{(a}Z_{b)}$, we have $h_{ab} = \phi n_a n_b$ where

$$\phi = \tfrac{1}{2}h''_{ab}x^a x^b = \tfrac{1}{2}w''_{ab}x^a x^b - \tfrac{1}{4}w'' x_a x^a, \tag{11.5}$$

with $w = w_a{}^a = -\frac{1}{2}\Box\phi$.

We can further refine the gauge of a plane wave, as follows. Suppose that the inertial coordinates have been chosen so that the spatial part of n is the unit vector in the z-direction. Then $u = t - z$ and the wave is travelling in the z-direction. Replace the t and z coordinates by $u = t - z$ and $v = t + z$. Then the Minkowski space metric becomes

$$\mathrm{d}u\,\mathrm{d}v - \mathrm{d}x^2 - \mathrm{d}y^2$$

and we have $x_a x^a = uv - x^2 - y^2$. The four-vector comonents n^a are $(0, 2, 0, 0)$. Note that lowering the index produces the covector n_a with components $(1, 0, 0, 0)$, with the nonzero component in the first, not the second position.

Let E, F, G denote, respectively, the xx, xy, and yy components of w_{ab} in the new coordinate system. Then $w = -E - G$. The quantities E, F, G, w are all functions of u alone. Because we also have $n^a w_{ab} = 0$, the term $\frac{1}{2} w''_{ab} x^a x^b$ on the right-hand side of (11.5) is independent of v. We can therefore write ϕ in the form

$$\phi = \psi + \chi'' v + \alpha'' x + \beta'' y + \gamma' , \tag{11.6}$$

where

$$\psi = \tfrac{1}{4}(E'' - G'')(x^2 - y^2) + F'' xy \tag{11.7}$$

and χ, α, β, γ are functions of the variable u alone. However

$$(\chi'' v + \alpha'' x + \beta'' y + \gamma') n_a n_b = 2\partial_{(a} W_{b)} ,$$

where

$$(W_a) = \tfrac{1}{2}(\chi' v + \alpha' x + \beta' y + \gamma, -\chi, -\alpha, -\beta) .$$

Therefore our metric disturbance is equivalent by a gauge transformation to one of the form $h_{ab} = \psi(u, x, y) n_a n_b$, where ψ is defined by (11.7). We call this the *null gauge*. We have

$$\Box \psi = 0, \qquad n^a \partial_a \psi = 0, \tag{11.8}$$

together with the condition that $\partial_a \partial_b \psi$ should be a function of $u = n_a x^a$ alone. Conversely, given ψ satisfying these conditions for some constant null vector n^a, the metric disturbance h_{ab} is equivalent to a plane wave.

Exercise 11.3

Show that if ψ satisfies the three conditions, then $h_{ab} = \psi n_a n_b$ is equivalent to a plane wave by a gauge transformation.

We therefore have the following alternative characterizations of a plane wave solution to the empty space linearized equations. They are equivalent by a gauge transformation.

– A metric disturbance h_{ab} for which, for some constant null vector n^a,

$$n^a h_{ab} = 0 \quad \text{and} \quad X^a \partial_a h_{bc} = 0$$

for every four-vector such that $X^a n_a = 0$.

– A metric disturbance of the form $h_{ab} = \psi n_a n_b$ for some constant null four-vector n^a, where

$$\Box \psi = 0 \quad \text{and} \quad n^a \partial_a \psi = 0 ,$$

and $\partial_a \partial_b \psi$ is a function of $u = n_a x^a$ alone.

In the second case, with an appropriate choice of coordinates, the disturbed Minkowski metric is

$$ds^2 = du\,dv - dx^2 - dy^2 + \psi(u,x,y)\,du^2\,. \tag{11.9}$$

This is a solution of the linearized Einstein equations whenever

$$\Box\psi = \frac{\partial^2\psi}{\partial x^2} + \frac{\partial^2\psi}{\partial y^2} = 0 \tag{11.10}$$

and it is a plane wave whenever ψ is a polynomial of degree two in x, y, with coefficients depending on u alone. It is a remarkable fact that (11.9) is also a solution to the full, not linearized, vacuum equations whenever $\psi(u,x,y)$ satisfies (11.10). Solutions of this form are called *pp-waves*. The 'pp' stands for 'plane-fronted with parallel rays', referring to the fact that the null four-vector with components $(0,1,0,0)$ is covariantly constant not only in the Minkowski background, but also with respect to the Levi-Civita connection of the disturbed metric.

Exercise 11.4

Show that the curvature tensor of a pp-wave satisfies

$$n^a R_{abcd} = 0\,.$$

In the linearized theory, this follows from the formula for the linearized curvature tensor (6.4). In the full theory, first establish that n^a is covariantly constant.

A plane wave can be detected through its curvature. A plane wave passing two particles in free-fall will produce a varying relative acceleration between them by the equation of geodesic deviation. Alternatively this will show up as a varying force between constrained particles. Early attempts at detection sought to observe the effect of this force in a large solid bar. Current attempts focus on the effect on the optical path lengths in what is essentially a large Michelson–Morley interferometer [13].

11.4 The Retarded Solution

We can understand the way in which Maxwell's equations describe the generation of electromagnetic waves by looking at the *retarded solution* to (11.1). This is

$$\Phi_a = \frac{k}{4\pi}\int J_a\,d\nu\,,$$

where the integral is over the past light-cone of the event at which Φ_a is evaluated and $d\nu$ is the invariant volume element on the light-cone; see [23], p. 148. If the event at which the potential is evaluated is (t', r'), then (t, r) lies on the past light-cone whenever the four-vector N with temporal and spatial parts $(t' - t, r' - r)$ is null and future pointing. We can use the components x, y, z of r as coordinates on the light-cone. Then $d\nu = dV/|r'-r|$, where $dV = dx\, dy\, dz$. The retarded solution becomes

$$\Phi_a(t', r') = \frac{k}{4\pi} \int |r - r'|^{-1}[J_a]\, dV\,,$$

where the integral is over r and the square brackets indicate *evaluation at retarded time*. That is, given a function $f(t, r)$ on space–time and the event (t', r'), we define

$$[f](r) = f(t' - |r' - r|, r)\,. \tag{11.11}$$

At a large distance from the source, the field looks like a combination of a Coulomb field, the field of a point charge Q, and electromagnetic waves. The value of Q is also given by an integral over the past light-cone of (t', r'):

$$Q = \int N^a J_a\, d\nu\,.$$

This is independent of t' and r', and is invariant under change of inertial coordinates. By the exercise below, the first statement is a consequence of the conservation law $\partial_a J^a = 0$; the second follows from the invariance of $d\nu$.

We do not derive here the asymptotic decomposition of the field into a Coulomb part and a radiation part because the theory is covered in many texts on electromagnetism. Instead, we look in detail at the less familiar decomposition of the linearized gravitational field of a bounded source, from which the electromagnetic theory can also be derived by analogy.

Exercise 11.5

Show that Q is independent of t' and r'.

By applying the same results to the weak field approximation to Einstein's equations for each value of b in turn, we can express the value of w_{ab} at each event as an integral over the past light-cone of the event:

$$w_{ab} = -4 \int T_{ab}\, d\nu\,.$$

Therefore it is the density and motion of the sources at events on the past light-cone that contribute to the metric perturbation at the event. We also have that the covector

$$p_a = \int N^b T_{ab}\, d\nu \tag{11.12}$$

is constant as a consequence of the conservation law $\partial_a T^{ab} = 0$. For physically reasonable matter, it is timelike and future-pointing. It represents the *four-momentum* of the source.

11.5 Quadrupole Moments

Before we explore further how changes in the source produce observable effects outside the source, we first look at how the approximation that we use works in the classical Newtonian theory. Here, with $G = 1$, we have

$$\nabla^2 \phi = 4\pi\rho \,,$$

where ϕ is the potential and ρ is the density of the source. For the gravitational field of a body enclosed in a volume V, this has solution

$$\phi(\mathbf{r}') = -\int_V |\mathbf{r} - \mathbf{r}'|^{-1} \rho(\mathbf{r}) \, \mathrm{d}V \,,$$

where $\mathbf{r}'$ is the position vector of the point at which ϕ is evaluated, and $\mathbf{r}$ is the position vector of a typical point of the body.

Consider the field a long way from the body. That is, assume that the origin is inside the body and that $r' = |\mathbf{r}'|$ is large compared to the dimensions of the body, and expand in inverse powers of r', discarding terms of order r'^{-4}. By using Taylor's formula, we have

$$
\begin{aligned}
\frac{1}{|\mathbf{r} - \mathbf{r}'|} &= r'^{-1}(1 + r'^{-2}\mathbf{r}.\mathbf{r} - 2r'^{-2}\mathbf{r}.\mathbf{r}')^{-1/2} \\
&= \frac{1}{r'} - \frac{\mathbf{r}.\mathbf{r} - 2\mathbf{r}.\mathbf{r}'}{2r'^3} + \frac{3(\mathbf{r}.\mathbf{r}')^2}{2r'^5} + O(r'^{-4}) \,.
\end{aligned}
\tag{11.13}
$$

Let us put $\mathbf{r}' = r'\mathbf{e}$, where $\mathbf{e}$ is the unit vector in the direction of $\mathbf{r}'$, and introduce the quantities

$$m = \int_V \rho \, \mathrm{d}V, \qquad c_i = \int_V \rho r_i \, \mathrm{d}V, \qquad q_{ij} = \int_V \rho(3r_i r_j - \delta_{ij} r_k r_k) \, \mathrm{d}V \,,$$

where the r_is are the components of $\mathbf{r}$ and there is summation for repeated indices over $1, 2, 3$. The quantity m is the total mass of the source, $\mathbf{c}/m$ is the position of the centre of mass, and the q_{ij}s are the *quadrupole moments* at the origin. If they are taken to be the entries in a matrix q, then

$$
q = (A + B + C)\begin{pmatrix} 1 & 0 & 0 \\ 0 & 1 & 0 \\ 0 & 0 & 1 \end{pmatrix} - 3\begin{pmatrix} A & -H & -G \\ -H & -B & -F \\ -G & -F & C \end{pmatrix} \,,
\tag{11.14}
$$

where A, B, C are the moments of inertia of the body at the origin and F, G, H are the products of inertia. That is,

$$A = \int_V \rho(y^2 + z^2)\, \mathrm{d}V, \qquad H = \int_V \rho xy\, \mathrm{d}V,$$

and so on. The second matrix on the right-hand side of (11.14) is the *inertia tensor* $\mathcal{J}$ of the body. Thus q is the trace-free part of $-3\mathcal{J}$. It vanishes for a body with spherical symmetry about the origin, and so can be seen as a measure of deviation from spherical symmetry.

With these definitions,

$$\phi(\boldsymbol{r}') = -\frac{m}{r'} - \frac{c_i e_i}{r'^2} - \frac{q_{ij} e_i e_j}{2r'^3} + O(r'^{-4}).$$

The first term is the potential of a point mass; the second vanishes if the origin is at the centre of mass. With $\boldsymbol{c} = 0$, the third term can be seen as a correction to the spherically symmetric field obtained by concentrating all the mass at the centre of mass. It shows the effect of irregularities in the distribution of matter in the source.

11.6 Generation of Gravitational Waves

We now apply a similar approximation to the retarded solution of the linearized Einstein equations to find out how the motion of matter within a source generates gravitational radiation. We choose the inertial coordinates so that the event at which w_{ab} is evaluated is $(t', \boldsymbol{r}')$, and so that the origin is inside the source.

Let V be a fixed volume containing the source, and denote by $\boldsymbol{r}$ the position vector in the inertial coordinates of a typical event happening within V. As in the Newtonian theory, the approximation is based on the assumption that $r' = |\boldsymbol{r}'|$ is large compared with the dimensions V; that is, we are considering the radiation field at a large distance from the source. We can write the retarded solution in the form

$$w_{ab}(t', \boldsymbol{r}') = -4 \int_V |\boldsymbol{r}' - \boldsymbol{r}|^{-1} [T_{ab}]\, \mathrm{d}V, \qquad (11.15)$$

where the square brackets indicate evaluation at retarded time (11.11). For large r', we have

$$w_{ab} = -\frac{4\tau_{ab}}{r'} + O(r'^{-2}),$$

where

$$\tau_{ab} = \int_V [T_{ab}]\, \mathrm{d}V$$

by using the approximation (11.13). Now τ_{ab} depends on t' and r' through the definition of the retarded time (11.11). However, by substituting

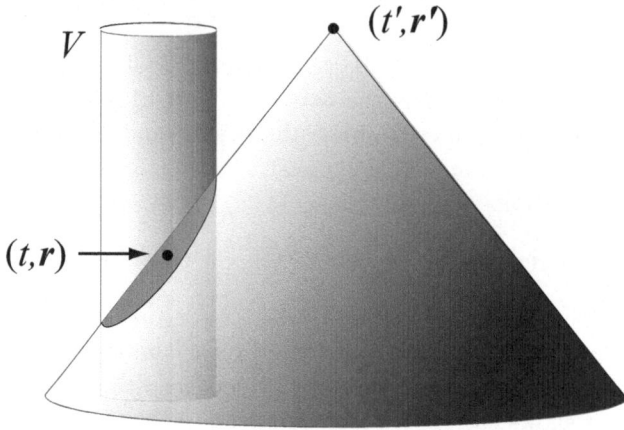

Figure 11.1 Evaluation of the retarded solution

$$\mathrm{d}\nu = \frac{\mathrm{d}V}{|r' - r|}$$

in (11.12), we find that

$$p_a = \int_V |r' - r|^{-1} N^b [T_{ab}] \, \mathrm{d}V$$

is constant, where N^a is the null four-vector $(|r' - r|, r' - r)$. As $r' \to \infty$

$$|r' - r|^{-1} N^a = n^a + O(r'^{-1}) \,,$$

where n is the null vector $(1, r'/r')$. Therefore

$$n^b \tau_{ab} = p_a + O(r'^{-1}) \,.$$

We assume, without loss of generality, that the inertial coordinates t, x, y, z have been chosen so that $p_a = m V_a$, where m is a constant, the mass of the source, and V^a is a four-velocity parallel to the t-axis, that is, so that the source is at rest in the inertial frame.

We need to separate w_{ab} at large distances into a part that we can identify as the static gravitational field associated with the total mass of the source and a second component that we can interpret as the radiation emitted by the source. To the leading order in r'^{-1}, the first will be a linearized Schwarzschild solution and the second will look like a plane wave moving directly away from the source.

The first step is to understand the dependence of τ_{ab} on the coordinates of the event at which the retarded solution is evaluated. Now we get the same change in the value of τ_{ab} by displacing the event (t', r') through a four-vector X^a as we get by displacing the source through $-X^a$. Therefore

$$\partial'_c \tau_{ab} = \int_V [\partial_c T_{ab}] \, \mathrm{d}V \,,$$

where ∂'_c and ∂_c are, respectively, the partial derivatives with respect to the inertial coordinates of the event (t', r') at which the metric disturbance is evaluated and the partial derivatives with respect to the coordinates of the event (t, r). By (A.2), we have

$$\int_V [\boldsymbol{\nabla} f] \, \mathrm{d}V = - \int_V [\partial_t f] e \, \mathrm{d}V \,,$$

where $e = (r' - r)/|r' - r|$. Therefore

$$\partial_c \tau_{ab} = \int_V |r' - r|^{-1} N_c [\partial_t T_{ab}] \, \mathrm{d}V \,.$$

So for large r', we have

$$\partial_c \tau_{ab} = n_c \int_V [\partial_t T_{ab}] \, \mathrm{d}V + O(r^{-1}) \,.$$

Now put $\sigma_{ab} = \tau_{ab} - m V_a V_b$. Then

$$n^a \sigma_{ab} = 0 \qquad X^c \partial_c \sigma_{ab} = 0$$

whenever $X^a n_a = 0$. We have

$$w_{ab} = -\frac{4m V_a V_b}{r'} - \frac{4m \sigma_{ab}}{r'} + O(r'^{-2}) \,.$$

The first term on the right is the (linearized) Schwarzschild solution for a mass m at rest. It is analogous to the 'Coulomb potential' in the electromagnetic case. In a neighbourhood of the point at which the field is evaluated at large r', we have

$$\frac{r'}{r'} = e + O(r'^{-1}) \,,$$

where e is a constant unit vector. Thus the second term looks like a plane wave travelling in the direction of e, directly away from the source.

We can relate the plane wave component to the derivatives of the quadrupole moments of the source, by putting $\rho = n^a n^b [T_{ab}]$ and by deriving the formula

$$\sigma_{ij} = \tau_{ij} = \frac{1}{2} \frac{\partial^2}{\partial t'^2} \int_V \rho r_i r_j \, \mathrm{d}V + O(r'^{-1}),$$

for the spatial components of σ_{ab} in the inertial coordinate system; here $i, j = 1, 2, 3$. To do this, we take r^a to be the four-vector with temporal and spatial parts $(0, \boldsymbol{r})$. Then with $(t', \boldsymbol{r'})$ fixed,

$$\partial_c \partial_d \big([T^{cd}] r^a r^b\big) = r^a r^b \partial_c \partial_d [T^{cd}] + 2\partial_c \big([T^{cd}] r^{(a} \partial_d r^{b)}\big) - 2[T^{cd}] \partial_c r^a \partial_d r^b.$$

Because $[T^{ab}]$ depends only on $\boldsymbol{r}$, the left-hand side is equal to

$$\partial_i \partial_j \big([T^{ij}] r^a r^b\big),$$

with summation over $i, j = 1, 2, 3$. Therefore the integral of the left-hand side over V vanishes by the divergence theorem. The integral of the second term on the right-hand side similarly vanishes. We then observe that $\partial_c r^a = 1$ if $c = a = 1, 2, 3$, and that it vanishes otherwise. So by taking $a = i$, $b = j$, we have

$$\int_V [T_{ij}] \, \mathrm{d}V = \tfrac{1}{2} \int_V r_j r_j \partial_c \partial_d [T^{cd}] \, \mathrm{d}V.$$

Finally, from (A.2) and the fact that $\partial_a T^{ab} = 0$, we have

$$\partial_a [T^{ab}] = N_a [\partial_t T^{ab}].$$

By applying this twice, we have $\partial_c \partial_d [T^{cd}] = N_c N_d [\partial_t^2 T^{cd}]$, and hence the required result.

From the discussion in §11.3, the radiation part of the metric disturbance is therefore gauge-equivalent to $\psi n_a n_b$, where

$$\begin{aligned} \psi &= -\frac{2}{r'} \frac{\mathrm{d}^2}{\mathrm{d}t'^2} \int_V \rho(x^2 - y^2 + 2xy) \, \mathrm{d}V + O(r'^{-2}) \\ &= r'^{-1} \big(2(\ddot{A} - \ddot{B})(x^2 - y^2) + 4\ddot{H}xy\big), \end{aligned}$$

where the dot denotes the derivative with respect to t' and A, B, H are the moments and products of inertia:

$$A = \int_V \rho(y^2 + z^2) \, \mathrm{d}V, \qquad B = \int_V \rho(x^2 + z^2) \, \mathrm{d}V, \qquad H = \int_V \rho xy \, \mathrm{d}V.$$

In this way the radiation field at large distances is determined by the second rates of charge of the moments and products of inertia along the axes orthogonal to the direction of the source. Because ψ depends only on the difference $A - B$, the radiation field is determined by the rates of change of the quadrupole moments.

12
Redshift and Horizons

When one observer sends light signals to another, the frequency of the light measured at emission by the first observer is generally not the same as that measured at reception by the second. Even in special relativity, the light is redshifted if the second is moving away from the first. This is the Doppler effect. In general relativity, there is a gravitational redshift when both are at rest in the gravitational field of a static spherically symmetric body, and the first is below the second.

In extreme cases the redshift becomes infinite when the first observer passes through an horizon. We saw this in the Schwarzschild solution when an observer falls through the event horizon. But the phenomenon can also occur in flat space–time, when the first observer is at rest and the second is accelerating uniformly. We consider this in Example (12.2) below.

In this chapter, we take a general look at the phenomenon of redshift, which is of great importance in cosmology, and at horizons. In particular, we consider briefly the 'horizon problem' in cosmology.

12.1 Retarded Time in Minkowski Space

Let O be an observer in Minkowksi space, with worldline ω. Then ω is a timelike curve, which we can parametrize by proper time τ, the time measured by a standard clock carried by O. In inertial coordinates, ω is given by

$$x^a = x^a(\tau),$$

and $V^a = \mathrm{d}x^a/\mathrm{d}\tau$ is a future-pointing timelike vector. We assume that ω is *complete* in the sense that τ extends from $-\infty$ to ∞ along ω.

Let $I^+(\omega)$ denote the set of events in Minkowski space that can be reached from an event on ω at less than the speed of light. That is, the set of events with coordinates y^a such that

$$T^a = y^a - x^a(\tau)$$

is future-pointing and timelike for some τ. This is called the *future set* of ω. Although it is an open subset, the second example below shows that $I^+(\omega)$ need not be the whole of Minkowksi space.

For any event E in $I^+(\omega)$ not on ω, there is a unique value of τ for which T^a is null and future-pointing. This value of τ is called the *retarded time* at E determined by ω (see Figure 12.1). If E is actually on ω, then the retarded

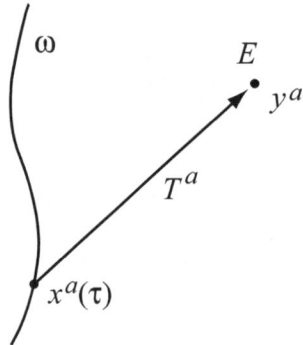

Figure 12.1 Retarded time

time is defined to be the proper time at E. We have already met one version of this definition in the last chapter in the context of finding the gravitational radiation emitted by a source. Radiation generated at an event on ω at proper time τ is seen by a second observer at an event with retarded time τ.

Exercise 12.1

Let E be the origin of the inertial coordinate system t, x, y, z and suppose that $E \in I^+(\omega)$. Show that there is a unique value of τ for which the four-vector from $x^a(\tau)$ to E is future-pointing and null.

We can similarly define the past set $I^-(\omega)$ and the advanced time at an event in $I^-(\omega)$ by substituting 'past-pointing' for 'future-pointing'.

Example 12.1

Suppose that O is at rest at the origin in an inertial coordinate system t, x, y, z. Then ω is given by

$$t = \tau, \qquad x = y = z = 0.$$

In this case, the future and past sets are the whole of Minkowski space. At a general event with coordinates t, x, y, z, the retarded time is $\tau = t - r$, where $r^2 = x^2 + y^2 + z^2$. The advanced time is $t + r$.

Example 12.2

Suppose that O has constant acceleration worldline

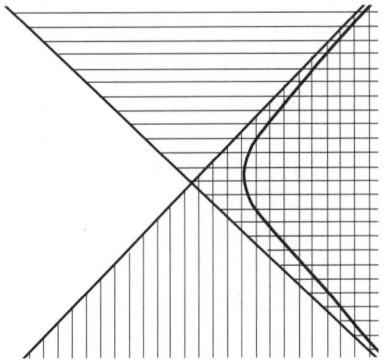

Figure 12.2   The future and past sets of an accelerating observer

$$t = \sinh(\tau), \qquad x = \cosh(\tau),$$

with unit acceleration, measured by O. In this case,

$$I^+(\omega) = \{t + x > 0\}, \qquad I^-(\omega) = \{t - x < 0\}.$$

In Figure 12.2, $I^+(\omega)$ is shaded horizontally and $I^-(\omega)$ is shaded vertically; the hyperbola is the worldline, and its asymptotes are $t = \pm x$. The retarded time at (t, x, y, z) goes to $-\infty$ as $t + x \to 0$.

12.2 Horizons

We can also define the future and past sets of a worldline ω in a general space–time provided that it is *time orientable*. That is, provided that it is possible to distinguish future-pointing from past-pointing timelike vectors continuously throughout space–time. The definitions are not quite as simple as in Minkowski space because there is no unambiguous notion of the displacement vector from one event to another. Instead we say that an event E lies in the *past set* $I^-(\omega)$ whenever there is a future-directed timelike curve from E to some event on ω. A future-directed timelike curve is a parametrized curve with future-pointing timelike tangent vector. The parameter must increase from E to ω. So $E \in I^-(\omega)$ if it is possible to travel from E to an event on ω at less than the speed of light. We similarly define the *future set* of ω by replacing 'future-directed' by 'past-directed'. We could equally well replace ω by any other subset of M, but our focus is on the future and past sets of observers' worldlines.

It can happen that $I^-(\omega)$ and $I^+(\omega)$ are both the whole of space–time, so the observer can influence any event in space–time and be influenced from it. In general, this will not be so. We have seen one example in Minkowski space. A second example is in the Schwarzschild metric in Eddington–Finkelstein coordinates. Here the past set of an observer at rest outside the horizon is the exterior of the black hole. The boundary of the past set is the boundary of the black hole.

In every case, however, the future and past sets are *open*. We do not prove this, but it is not hard to do so. The key idea is that if γ is a future-directed timelike curve from E to an event on ω, then it is possible to perturb γ and move E in a neighbourhood of E while keeping γ timelike.

The boundary of $I^-(\omega)$, in the topological sense, is called the observer's *event horizon*. It is important to realise that in a general space–time, the 'event horizon' is something that depends on the observer. It need not be a smooth hypersurface, but when it is, it must be null. In fact if f is a smooth function on some neighbourhood in space–time with nonvanishing gradient and with the property that $f(E) < 0$ if $E \in I^-(\omega)$ and $f(E) \geq 0$ if $E \notin I^-(\omega)$, then $\nabla_a E$ is future-pointing and null on the boundary where $f = 0$.

Exercise 12.2

Show that if Σ is given by $f = 0$ and if $n^a = \nabla^a f$ is null and future-pointing at every $E \in \Sigma$, then there is a null geodesic contained in Σ through every event in Σ.

Thus a smooth event horizon is ruled by null geodesics, as in the case of the two examples. Penrose shows that this is true more generally in [19].

A more interesting example is the Kerr space–time (10.9), which models the gravitational field outside a rotating body. It is a stationary space–time in the sense that it admits a timelike Killing vector: the vector field t^a with components $(1, 0, 0, 0)$ in Boyer–Lindquist coordinates t, r, θ, φ is a Killing vector and is timelike, at least for large r. One can therefore pick out 'observers at rest' in the region in which t^a is timelike by the condition that they should have four-velocities parallel to t^a, or equivalently by the condition that r, θ, φ should be constant on their worldlines.

Consider such observers in the region $r > r_0$, for some large value of r_0. Because the metric here is close to that of flat space–time, it is reasonably clear that causal relations between such obervers should be the same as in Minkowski space.[1] The whole of the region $r > r_0$ should be in the past of any one of them. In other other words, any event happening at $r > r_0$ should be visible to every stationary observer at $r > r_0$.

Exercise 12.3

Show that for a stationary observer in the Kerr space–time at $r > r_0$ for large r_0, the whole of the region $r > r_0$ is in the past set of ω.

So what is the full extent of the past set of a stationary observer at large r? The gradient covector $\nabla_a r$ has components $(0, 1, 0, 0)$ in Boyer–Lindquist coordinates, and therefore from (10.9), we have

$$\nabla_a r \nabla^a r = -\frac{\Delta}{\Sigma} = -\frac{r^2 + a^2 - 2mr}{r^2 + a^2 \cos^2 \theta} \,.$$

Thus $\nabla_a r$ is spacelike whenever $r > r_+$, where r_+ is the larger root of

$$r^2 + a^2 - 2mr \,.$$

At any event at which $\nabla_a r$ is spacelike, it is possible to find a future-pointing timelike vector T^a such that $T^a r_a$ is positive, so that r is increasing along T^a. It follows that we can construct a future-directed timelike curve from any event in the region $r > r_+$ to the region at large r. Therefore that for any stationary observer at large r with worldline ω is

$$I^-(\omega) \supseteq \{r > r_+\} \,.$$

In Boyer–Lindquist coordinates, the metric coefficients are singular at r_+. As in the Schwarzschild metric, however, this is simply an artefact of the co-ordinate choice. In the coordinate system (10.10), the singularity disappears, and $\nabla_a r$ becomes a past-point null vector at $r = r_+$. Therefore $T^a \nabla_a r < 0$ at $r = r_+$ for any future-pointing timelike T^a and so r is decreasing along

---

[1] It is not true, however, that the causal relations between events are the same as in Minkowski space.

any future-directed timelike curve at $r = r_+$. It follows that none of the region $r < r_+$ in the coordinate system (10.10) can be in the past of a distant stationary observer.

We conclude that for a distant stationary observer with complete worldline,

$$I^-(\omega) = \{r > r_+\}.$$

The null hypersurface $r = r_+$ is the common event horizon of such observers. It is the boundary of the rotating black hole represented by the Kerr metric.

Several points should be noted here. The first is that although all distant stationary observers at rest share the same event horizon, it is not true that all noninertial observers have this horizon; it is not even true in Minkowksi space that all observers have the same horizon. Second, the problem of characterizing the boundary of a black hole in a general dynamical setting, in which the metric is not stationary, is nontrivial. Third, as in the Schwarzschild solution, there is an alternative extension in which the metric represents a 'white hole', and a larger extension with two exterior regions joined by a wormhole. In fact the maximally extended Kerr space–time contains an infinite number of 'exterior regions'. It also contains closed timelike curves, which violate causality [8]. Fourth, in contrast to the Schwarzschild case, the event horizon in the Kerr space–time is not the same as the surface of 'infinite redshift'.

To expand on this last remark, suppose that we have two stationary observers at r_1, θ_1, φ_1 and r_2, θ_2, φ_2 in the Boyer–Lindquist coordinates. If the first sends a photon with frequency ω_1, then by the same argument as in §7.5, it will be seen by the second to have frequency

$$\omega_2 = \omega_1 \sqrt{\frac{g_{00}(r_1)}{g_{00}(r_2)}} = \omega_1 \sqrt{\frac{(r_1^2 + a^2 \cos^2 \theta_1 - 2mr_1)(r_2^2 + a^2 \cos^2 \theta_2)}{(r_2^2 + a^2 \cos^2 \theta_2 - 2mr_2)(r_1^2 + a^2 \cos^2 \theta_1)}}.$$

The frequency ω_2 goes to zero, that is, the redshift becomes infinite, when

$$r_1^2 + a^2 \cos^2 \theta_1 - 2mr_1 \to 0.$$

In fact if the left-hand side becomes negative, then t^a is no longer timelike, and there are no stationary observers. There are, however, events outside the event horizon at which t^a is spacelike. These make up the so-called *ergosphere* of the black hole. As a stationary source of light is moved (slowly) towards the ergosphere, any light it emits becomes infinitely redshifted as it approaches the boundary of the ergosphere, well before it reaches the event horizon.

Penrose [17] observed that it is possible in principle to extract rotational energy from a rotating black hole. The quantity $E = V_a t^a$ is conserved along the worldline of a unit mass particle with four-velocity V^a. It is the total energy of the particle, including its rest energy. It is also conserved in collisions.

In a region in which l^a is spacelike, it is possible for E to be negative for some timelike four-velocities V^a. One can imagine a particle falling into the ergosphere, and splitting into two pieces, one of which has $E < 0$. The piece with negative E must fall into the black hole, and cannot escape back to infinity. The other piece, however, will have a larger value of E than the original particle and can escape back to infinity. This second fragment gains energy at the expense of the rotational energy of the black hole itself.

12.3 Homogeneous and Isotropic Metrics

A homogeneous and isotropic cosmology is one that looks the same everywhere and in every direction. Its properties extend the Copernican principle that the earth should not be seen as occupying a central place in the universe. There is a small class of such cosmological models that are also homogeneous in time, and we look at these briefly below. But a general homogeneous and isotropic space–time is not static, and so we can use the geometry to pick out a universal time coordinate, for example, by taking t to be the scalar curvature. Its gradient $t^a = \nabla^a t$ is a natural vector field, which we assume to be timelike, so that it everywhere determines a standard of rest. Homogeneity and isotropy are then the requirements that the universe should look the same everywhere at any given time to an observer at rest, and in every direction. We now derive the most general metric with these properties, and find its Ricci curvature so that we can determine its dynamical behaviour from Einstein's equation.

An immediate consequence of the requirements is that $t^a t_a$ must be a function of t alone. By replacing t by a function of t, we can set $t^a t_a = 1$. Then t is the proper time of an observer at rest. Such an observer must be in free-fall, as one can see either from isotropy—acceleration would give a preferred direction—or from

$$t^a \nabla_a t_b = t^a \nabla_a \nabla_b t = t^a \nabla_b \nabla_a t = \tfrac{1}{2} \nabla_b (t^a t_a) = 0 \,. \tag{12.1}$$

A second consequence is that the Ricci tensor must be of the form

$$R_{ab} = \mu t_a t_b + \lambda g_{ab} \,, \tag{12.2}$$

for some scalars λ and μ, otherwise its eigenvectors, the solutions to

$$R_{ab} V^a \propto V_b \,,$$

would pick out preferred directions in space. The scalars must be functions of t alone, by homogeneity. For the same reason, and because $t^a \nabla_a t_b = 0$,

$$\nabla_a t_b = \beta(g_{ab} - t_a t_b) \tag{12.3}$$

for some function β of t. It follows that $\nabla_a t^a = 3\beta$. However, from the definition of the curvature tensor and (12.1),

$$
\begin{aligned}
3\dot\beta = t^a \nabla_a \nabla_b t^b &= t^a \nabla_b \nabla_a t^b + t^a t^b R_{ab} \\
&= \nabla_b(t^a \nabla_a t^b) - (\nabla_b t^a)\nabla_a t^b + t^a t^b R_{ab} \\
&= \mu + \lambda - 3\beta^2 \,,
\end{aligned}
$$

where the dot denotes the derivative with respect to t. Now introduce coordinates x^1, x^2, x^3 to label the worldlines of the observers at rest, and put $x^0 = t$. Then the metric must take the form

$$
ds^2 = dt^2 + g_{ij}dx^i dx^j \,, \tag{12.4}
$$

with summation over $i, j = 1, 2, 3$. There can be no $dt\,dx^i$ terms or the corresponding metric coefficients g_{0i} would determine a preferred direction in space. By Exercise 7.2, we have $\partial_t g_{ab} = 2\nabla_{(a} t_{b)}$ in these coordinates, and thus

$$
\partial_t g_{ij} = 2\beta\, g_{ij} \,. \tag{12.5}
$$

Concentrate now on one observer, whom we take to be at the origin of the spatial coordinates x^i, and consider in more detail the consequences of the isotropy assumption, which implies that the metric should be spherically symmetric about the observer's location. As in our derivation of the Schwarzschild space–time, together with (12.5), this implies that the observer should be able to pick the spatial coordinates to be $x^1 = r$, $x^2 = \theta$, $x^3 = \varphi$, so that

$$
g_{ij}dx^i dx^j = -\alpha^2 \big(B\,dr^2 + Cr^2(d\theta^2 + \sin^2\theta\,d\varphi^2)\big)\,,
$$

where B, C are functions of r, and α is a positive function of t, related to β by $\beta = \dot\alpha/\alpha$. As in our earlier analysis, we can set $C = 1$ by making a change in the r-coordinate.

Proposition 12.3

A static, homogeneous, and isotropic cosmology must have *Robertson–Walker metric*

$$
ds^2 = dt^2 - \alpha^2\big((1 - kr^2)^{-1}\,dr^2 - r^2(d\theta^2 + \sin^2\theta\,d\varphi^2)\big) \tag{12.6}
$$

and Ricci tensor

$$
R_{ab} = 3\alpha^{-1}\ddot\alpha t_a t_b + (\alpha^{-1}\ddot\alpha + 2\alpha^{-2}\dot\alpha^2 + 2k\alpha^{-2})(g_{ab} - t_a t_b)\,,
$$

where where k is constant and α is a function of t.

Proof

A further coordinate change

$$x = r \sin\theta \, \cos\varphi, \qquad y = r \sin\theta \, \sin\varphi, \qquad z = r \cos\theta \,,$$

brings the space–time metric (12.4) into the form

$$ds^2 = dt^2 - \alpha^2 \left(dx^2 + dy^2 + dz^2 + (B-1)dr^2 \right), \tag{12.7}$$

with r now defined by $r^2 = x^2 + y^2 + z^2$. There are three Killing vectors X, Y, Z, with respective components

$$(X^a) = (0, 0, -z, y)), \qquad (Y^a) = (0, z, 0, -x), \qquad (Z^a) = (0, -y, x, 0),$$

the last having components $(0, 0, 0, 1)$ in the t, r, θ, φ system. In the (x, y, z) coordinates, they correspond to the symmetries under rotations about the x, y, z axes, respectively.

By Exercise 5.12, we have

$$\nabla_a \nabla_b X_c = R_{bcad} X^d$$

and hence

$$
\begin{aligned}
X^c \nabla_a \nabla^a X_c &= \tfrac{1}{2}\Box(X_c X^c) - (\nabla_a X_c)(\nabla^a X^c) \\
&= R_{cd} X^c X^d \\
&= \lambda X_c X^c \,,
\end{aligned}
$$

where λ is as in (12.2) and $\Box = \nabla_a \nabla^a$ is the wave operator. With the metric given by (12.7), the components of the covector X_a are

$$\alpha^2 (0, 0, z, -y) \,.$$

The contravariant metric tensor is

$$g^{ab} = t^a t^b + \alpha^{-2}(t^a t^b - \delta^{ab} + kE^a E^b) \,,$$

where $(E^a) = (0, x, y, z)$ and k is defined in terms of B by

$$B = \frac{1}{1 - kr^2} \,.$$

It follows that

$$X_a X^a = -\alpha^2 (y^2 + z^2) \,.$$

Moreover $\nabla_a X_b = \partial_{[a} X_{b]}$ because $\nabla_a X_b$ is skew-symmetric and because the Levi-Civita connection is torsion-free. Therefore

$$
\begin{aligned}
(\nabla_a X_c)(\nabla^a X^c) &= g^{ac} g^{bd} \partial_{[a} X_{b]} \partial_{[c} X_{d]} \\
&= -2(k + \dot{\alpha}^2)(y^2 + z^2) + 2 \,.
\end{aligned}
$$

We have similar equations for the other two Killing vectors. By combining them, we get

$$-\Box(\alpha^2 r^2) = -2\lambda\alpha^2 r^2 - 4(k + \dot{\alpha}^2)r^2 + 6 \,.$$

But the wave operator is given by (5.1), with

$$|g| = \alpha^6 B r^4 \sin^2\theta$$

in the coordinates t, r, θ, φ. Therefore

$$\Box\alpha^2 = 2\alpha\ddot{\alpha} + 8\dot{\alpha}^2, \quad \alpha^2\Box r^2 = -6 + 8kr^2 + k'r^3 \,,$$

where the dot is the derivative with respect to t and the prime is the derivative with respect to r. Also,

$$\Box(\alpha^2 r^2) = r^2\Box\alpha^2 + \alpha^2\Box r^2$$

because the gradients of r and α are orthogonal. Finally, therefore,

$$\lambda = \alpha^{-1}\ddot{\alpha} + 2\alpha^{-2}\dot{\alpha}^2 + \tfrac{1}{2}\alpha^{-2}(4k + r^{-1}k') \,.$$

Because α is a function of t alone and k is a function of r alone, we must have that $4k + r^{-1}k'$ is constant. This implies that either k is constant or that it is a constant multiple of r^{-4}. The latter is not possible because it would make the metric singular at $r = 0$. So k is constant and the proposition follows. $\square$

The metric is unchanged if we replace r, α, and k by κr, α/κ, and k/κ, respectively, for some some positive constant κ. There is therefore no loss of generality in requiring that k should be one of $0, 1, -1$. In the first case, the spatial metric

$$\alpha^2(\mathrm{d}r^2 + r^2\mathrm{d}\theta^2 + r^2\sin^2\theta\,\mathrm{d}\varphi^2)$$

at a given time is simply a multiple of the metric on Euclidean space. In the case $k = 1$, the spatial metric is

$$\alpha^2\big(\mathrm{d}\chi^2 + \sin^2\chi\,\mathrm{d}\theta^2 + \sin^2\chi\sin^2\theta\,\mathrm{d}\varphi^2\big)$$

where $r = \sin\chi$. The expression in brackets is the metric on the hypersphere

$$w^2 + x^2 + y^2 + z^2 = 1$$

in $\mathbb{R}^4$, written in hyperspherical coordinates

$$x = \cos\varphi\sin\theta\sin\chi, \quad y = \sin\varphi\sin\theta\sin\chi, \quad z = \cos\theta\sin\chi, \quad w = \cos\chi.$$

Any point on the hypersphere can be taken as the origin $\chi = 0$, so all points in space are on the same footing. The spatial metric really is homogeneous.

In the case $k = -1$, the spatial metric is

$$\alpha^2\big(\mathrm{d}\chi^2 + \sinh^2\chi(\mathrm{d}\theta^2 + \sin^2\theta\,\mathrm{d}\varphi^2)\big) \,,$$

where now $r = \sinh\chi$.

Exercise 12.4

Show that in the case $k = -1$ the spatial metric is a multiple of the metric on the unit hyperboloid $t^2 - x^2 - y^2 - z^2 = 1$ in Minkowski space. Hence complete the argument that a homogeneous cosmology with $k > 0$ is *closed* in the sense that the hypersurfaces of constant t have the topology of the three-sphere, whereas those with $k \leq 0$ are open, with spatial topology $\mathbb{R}^3$.

The scalar curvature of the metric is

$$R = g^{ab}R_{ab} = 6(\alpha^{-1}\ddot{\alpha} + \alpha^{-2}\dot{\alpha}^2 + k\alpha^{-2})$$

and therefore the Einstein tensor is

$$\begin{aligned} G_{ab} &= R_{ab} - \tfrac{1}{2}Rg_{ab} \\ &= -3\alpha^{-2}(\dot{\alpha}^2 + k)t_a t_b - \alpha^{-2}(2\alpha\ddot{\alpha} + \dot{\alpha}^2 + k)(g_{ab} - t_a t_b) \,. \end{aligned}$$

Thus the energy-momentum tensor must be of the same form as that of a fluid. If our cosmology is to be interpreted as a solution of Einstein's equations, then it must be filled with fluid with density and pressure

$$\rho = \frac{3(k + \dot{\alpha}^2)}{8\pi\alpha^2}, \qquad p = -\frac{2\alpha\ddot{\alpha} + \dot{\alpha}^2 + k}{8\pi\alpha^2} \tag{12.8}$$

and four-velocity t^a. All that is needed to construct a model universe is to specify the relationship between ρ and p. That is, to choose an equation of state or equivalently to make some assumption about the physical nature of the matter filling the universe. We can then obtain from these two equations a single differential equation for α, and hence determine the evolution of the space–time geometry.

The function $\alpha(t)$ is called the *scale factor*. As α increases, the distance between points with fixed spatial coordinates increases in proportion, and the universe 'expands', although, as always, such a statement needs careful interpretation in terms of observations. In this context, it means no more than that the distance between two nearby observers at rest, as measured by either, is proportional to α.

If we combine the two equations to eliminate $\dot{\alpha}$, then we get

$$\frac{\ddot{\alpha}}{\alpha} = -\frac{4\pi(3p + \rho)}{3} \,.$$

On any conventional assumption about the nature of the matter filling the universe, ρ will be positive. Provided that $3p + \rho$ is positive, that is, provided that the pressure is not large and negative, we shall have $\ddot{\alpha} < 0$ throughout the history of the universe. From this, we can deduce the following 'singularity theorem'.

Proposition 12.4

Suppose that $\dot{\alpha}(t) > 0$ at some time t and that the energy condition $\rho + 3p \geq 0$ holds at all times. Then $\alpha(t_0) = 0$ for some $t_0 < t$.

Proof

Because $\ddot{\alpha} < 0$, Taylor's theorem with remainder implies that

$$\alpha(t') \leq \alpha(t) + \dot{\alpha}(t)(t' - t)$$

for all t'. The right-hand side vanishes when

$$t' - t = -\alpha(t)/\dot{\alpha}(t) < 0 \,,$$

therefore the left-hand side must also vanish for some $t_0 < t$.   □

In other words, if the universe is expanding at time t and is filled with matter with reasonable physical properties, then there must be a time in the past when the scale factor vanishes and at which the metric is therefore singular. The singularity is the 'big bang' of modern cosmology. The inequality $\rho + 3p \geq 0$ is part of the *strong energy condition*, which requires that

$$\rho + p > 0 \quad \text{and} \quad \rho + 3p > 0 \,.$$

By extending the methods of differential topology introduced into relativity by Penrose, Hawking and Penrose proved versions of this singularity theorem from the strong energy condition and other similar conditions under very general circumstances, without assuming homogeneity and isotropy; see [8]. Thus the existence of the initial singularity is a general consequence of Einstein's equation, and is not simply an artificial consequence of assuming a high degree of symmetry.

12.4 Cosmological Models

A simple choice for equation of state is $p = \nu\rho$, where ν is constant. If the dominant form of matter is galaxies, then it is reasonable to take the pressure to be zero, so that $\nu = 0$. If matter is dominated by radiation, then we would take $\nu = \frac{1}{3}$ because $T^a_{\ a} = 0$ for an electromagnetic energy-momentum tensor.

On eliminating ρ and p between the two equations, (12.8) gives

$$2\alpha\ddot{\alpha} + n(\dot{\alpha}^2 + k) = 0 \,, \tag{12.9}$$

where $n = 1 + 3\nu$. The energy condition $\rho + 3p > 0$ in the singularity theorem is $n > 0$. By writing $2\ddot{\alpha} = \mathrm{d}\dot{\alpha}^2/\mathrm{d}\alpha$ and integrating with respect to α, we have

$$\dot{\alpha}^2 + k = C\alpha^{-n} \,,$$

where C is constant.

We can see, in qualitative terms, the overall history of the universe by sketching the curves in the $\alpha, \dot{\alpha}$-plane determined by this equation for different values of the constant. The result is shown in Figure 12.3. If $k \leq 0$, then the universe expands from the initial singularity at which $\alpha = 0$ and $\dot{\alpha}$ is infinite; as $t \to \infty$, we have $\alpha \to \infty$ and $\dot{\alpha} \to \sqrt{-k}$, thus the initial expansion continues without limit. If on the other hand $k > 0$, then the expansion reaches a maximum before the universe recollapses to a final singularity.

By looking for solutions with the asymptotic behaviour $\alpha = O\big((t - t_0)^\sigma\big)$ as $t \to t_0$, one can also see from (12.9) that

$$\alpha = O\big((t - t_0)^{2/(n+2)}\big) \tag{12.10}$$

as $t \to t_0$. This gives the behaviour of α near the initial singularity in all the cases, and also near the final singularity in the closed case.

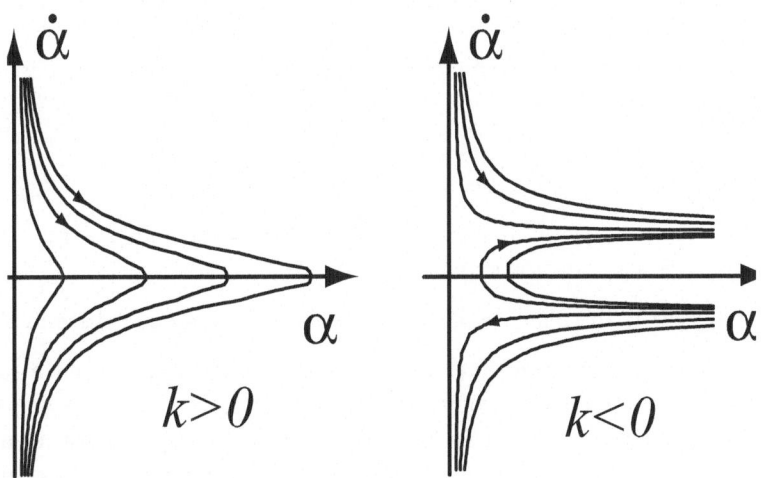

Figure 12.3 Phase portrait of the universe

Exercise 12.5

If the equation of state is $p = f(\rho)$, where $f > -1/3$, show that the phase curves are given by

$$g(\dot{\alpha}^2 + k) = C\alpha^{-1}, \qquad g(x) = \exp\left(-\frac{1}{3}\int \frac{\mathrm{d}x}{x + f(x)}\right).$$

Exercise 12.6

Show that in the case $\nu = 0$ (matter domination), $\rho\alpha^3$ is constant. Hence find α explicitly and verify the deductions from the phase portrait in the three cases $k < 0$, $k = 0$, and $k > 0$. Show also that ρ becomes unbounded at the initial singularity.

12.5 Homogeneity in Time

We derived the Robertson–Walker metric on the assumption that the scalar curvature was not constant, as well as the assumptions of isotropy and spatial homogeneity. There are two interesting cases in which the Robertson–Walker metric does have constant scalar curvature, and in which the space–time is also homogeneous in time.

The scalar curvature is constant for the Robertson–Walker metric whenever

$$\alpha^{-1}\ddot{\alpha} + \alpha^{-2}(\dot{\alpha}^2 + k)$$

is constant. The two obvious possibilities are the following.

Einstein static universe. In this case, $k > 0$, $\dot{\alpha} = 0$, and $\rho + 3p = 0$. The universe is closed, but not expanding.

de Sitter metric. In this case, $k = 0$ and $\alpha = \mathrm{e}^{Ht}$ for some constant H.

In neither case does our energy condition hold, so the metric is not a solution of Einstein's equation with a conventional form of matter as the gravitational source. In the first case, Einstein evaded this problem by suggesting a modification of the equations, which he later greatly regretted, by introducing a *cosmological constant* Λ. The modified equation is

$$R_{ab} - \tfrac{1}{2}Rg_{ab} - \Lambda g_{ab} = -8\pi T_{ab}.$$

Because $\nabla^a g_{ab} = 0$, this is consistent with the conservation equation $\nabla_a T^{ab} = 0$. If we take

$$T^{ab} = \rho t^a t^b, \qquad \Lambda = k/\alpha^2,$$

then the Einstein static universe is a solution with constant ρ and α. It is thus a static, dust-filled cosmology, but governed by a revised form of the field equation. Equivalently, one can take the term Λg_{ab} to the right-hand side and see the modification as the addition of a uniform distribution of 'matter' with constant unphysical density and pressure. The first interpretation fell from grace with Hubble's observation of the redshifts of galaxies, which implied that the universe is expanding and not static. Einstein had missed the opportunity to predict the expansion by rejecting the nonstatic solutions in favour of an inelegant tinkering with the original field equation.

In the second case, the metric is the *de Sitter* metric

$$\mathrm{d}t^2 - \mathrm{e}^{2Ht}(\mathrm{d}r^2 + r^2\,\mathrm{d}\theta^2 + r^2\sin^2\theta\,\mathrm{d}\varphi^2)\,. \tag{12.11}$$

This is homogeneous in space, and isotropic, but appears not to be static. But in fact it has a much larger symmetry group than the other Robertson–Walker metrics.

With $H = 1$, this can be seen by mapping the de Sitter space–time onto part of the spacelike hyperboloid

$$v^2 - w^2 - x^2 - y^2 - z^2 + 1 = 0$$

in the five-dimensional Minkowski space with metric

$$\mathrm{d}v^2 - \mathrm{d}w^2 - \mathrm{d}x^2 - \mathrm{d}y^2 - \mathrm{d}z^2\,.$$

The map is given by

$$\begin{aligned}
v + w &= \mathrm{e}^t \\
v - w &= r^2\mathrm{e}^t - \mathrm{e}^{-t} \\
x &= r\mathrm{e}^t \cos\varphi \sin\theta \\
y &= r\mathrm{e}^t \sin\varphi \sin\theta \\
z &= r\mathrm{e}^t \cos\theta\,.
\end{aligned}$$

The Minkowksi metric and the hyperboloid are invariant under the 'Lorentz group' of the five-dimensional space. This group has ten independent generators, so de Sitter space–time has the same number of independent Killing vectors as Minkowski space.

Exercise 12.7

Show that the de Sitter space–time is mapped onto the part of the hyperboloid $v + w > 0$. Show that the Minkowski metric coincides with the de Sitter metric on the hyperboloid.

The de Sitter space–time was the model for the *steady-state* cosmology, which was derived from the principle that, in an appropriate frame, the universe should look the same at all events. This is true of the hyperboloid because, given any two points on it, there is an element of the Lorentz group of the five-dimensional space which maps one to the other. It is thus a model universe which is homogeneous in time and space. In the steady-state theory, the galaxies were at rest in our original coordinates, but their density remained constant through the continuous creation of matter as the universe expanded.

Like the Einstein static model, the steady-state theory fails because it has no 'big bang', but the metric itself still plays a role in the context of 'inflation', which is discussed below. It is a solution of Einstein's equations with $p = -\rho$.

12.6 Cosmological Redshift

In the case of the Schwarzschild metric, we derived a 'redshift formula' relating the frequencies of light emitted and received by observers at rest by exploiting the existence of a timelike Killing vector. A general Robertson–Walker is not stationary, and the timelike vector field t^a that determines the standard of rest at each event is not a Killing vector. It is, however, a scalar multiple of a *conformal Killing vector*, that is, a vector field T^a that satisfies

$$\nabla_{(a} T_{b)} \propto g_{ab}. \tag{12.12}$$

In fact from (12.3), we have

$$\nabla_a t_b = 2\alpha^{-1}\dot{\alpha}(g_{ab} - t_a t_b)$$

and hence $\nabla_a(\alpha t_b) = \alpha g_{ab}$ because $\nabla_a \alpha = \dot{\alpha} t_a$. Therefore $T^a = \alpha t^a$ is a conformal Killing vector.

Now suppose that K^a is the tangent to the null geodesic worldline of a photon, and that the geodesic has affine parameter σ. Then

$$K^a \nabla_a K_b = 0, \qquad g_{ab} K^a K^b = 0$$

and therefore

$$\frac{\mathrm{d}}{\mathrm{d}\sigma}(T_b K^b) = K^a \nabla_a(T_b K^b) = K^a K^b \nabla_a T_b = K^a K^b \nabla_{(a} T_{b)} = 0.$$

So $T_a K^a$ is constant along the geodesic.

An observer at rest has four-velocity t^a. So the frequency ω measured by such an observer at an event on the geodesic is

$$\omega = t_a K^a = \alpha^{-1} T_a K^a.$$

It follows that if the photon is emitted at an event E_1 with frequency ω_1, as measured at E_1 by an observer at rest, and is received by a second observer at rest at an event E_2, then the second observer will measure a frequency ω_2, where

$$\alpha(E_1)\omega_1 = \alpha(E_2)\omega_2 .$$

In an expanding universe, α is an increasing function of t, and therefore the frequency measured by an observer at rest decreases with time. This is the *cosmological redshift*. It provides an explanation of what is sometimes referred to as one of the few unambiguous observations in cosmology, that the sky at night is dark. In an infinite, homogeneous, nonexpanding universe, it would be very bright. For although the intensity of light reaching us from an individual star falls off with the square of its distance, the total number of stars at a given distance increases in the same proportion. So the total intensity of the light reaching us from the stars is unbounded. This is *Olber's paradox*. It is resolved if the photons from the more distant stars are redshifted, and therefore their energy reduced, by the expansion of the universe.

The frequency measured by an observer at rest decreases along the worldline of a photon according to

$$\frac{\dot{\omega}}{\omega} = -\frac{\dot{\alpha}}{\alpha} .$$

The quantity $\dot{\alpha}/\alpha$ is called the *Hubble constant*, and is denoted by H. The terminology is potentially confusing because although the value of H is constant over space, it varies with t.

Suppose that light reaches us from a nearby galaxy, that it was emitted with known frequency ω and wavelength $\lambda = \omega^{-1}$, and that its wavelength on arrival is $\lambda + \delta\lambda$. The *redshift* is defined to be

$$z = \delta\lambda/\lambda .$$

Because the distance to a nearby galaxy, measured by an observer at rest, is proportional to the time that light takes to travel from the galaxy, we have that z increases in proportion to distance, at least for small distances. One can measure z by examining the shift in known spectral lines. So if one has some means of estimating the distance to the galaxy, one can measure H, and hence the current rate of expansion of the universe.

The best current measurement is that H^{-1} is 14 billion years. Knowledge of H allows us to relate the current value of ρ to k by

$$\frac{k}{\alpha^2} = \frac{8\pi\rho}{3} - H^2 .$$

Whether the universe is open ($k \leq 0$) or closed ($k > 0$) depends on the relative magnitudes of ρ and the *critical density* $3H^2/8\pi$. By the argument in the proof of the singularity theorem, Proposition 12.4, H^{-1} is an upper bound on the age of the universe, provided that the energy condition holds.

Exercise 12.8

Show that if T^a is a nonvanishing vector field and if coordinates are chosen so that its components are $(1, 0, 0, 0)$, then the condition (12.12) is $\partial_0 g_{ab} \propto g_{ab}$. A conformal Killing vector is therefore associated with a symmetry of the metric up to an overall scale. Show also that T^a is a conformal Killing vector if and only if $\nabla_{(a} T_{b)} = \frac{1}{4} \nabla_c T^c \, g_{ab}$.

12.7 Cosmological Horizons

In cosmology, we do not have the luxury of waiting indefinitely to see whether predictions about the future of the universe turn out to be correct. In considering the causal properties of a cosmological space–time, we are less interested than in the black hole case in which parts can be seen by an observer with a complete worldline, because it is not sensible to think in terms of observations made by an observer who will survive over cosmological timescales. It is more productive to ask questions about what can be deduced about the past history of the universe, and in particular about the behaviour of matter in the extreme conditions near the initial singularity, from observations made at the present. So we are more interested in determining the region of space–time that can be influenced by events on the worldline of a piece of matter at rest than on determining which events might be visible, eventually, to a stationary observer.

In the Robertson–Walker metric, the null geodesics passing through an event at $r = 0$ have constant θ and φ, by isotropy. Because they are null, they are therefore given by

$$\alpha \frac{\mathrm{d}r}{\mathrm{d}t} = \pm \sqrt{1 - kr^2} \, .$$

The plus sign gives the worldlines of photons emitted at $r = 0$. Thus a photon emitted at $r = 0$ at time t_0 can be seen at event (t, r, θ, φ) if

$$\int_{t_0}^t \frac{\mathrm{d}t'}{\alpha(t')} = \int_0^r \frac{\mathrm{d}r'}{\sqrt{1 - kr'^2}} \, . \tag{12.13}$$

If we take the big bang singularity to be at t_0, then this equation determines r as a function of t. It determines what is called the *particle horizon* of the worldline at $r = 0$ at time t. This is the surface at time t that separates particles that can have been influenced by events on the worldline at $r = 0$ since the big bang from those that cannot have been.

It appears at first that we have defined the particle horizon only for the worldline at $r = 0$. But because the metric is spatially homogeneous, the worldline can be that of any particle at rest. We can rewrite the formula in a way that makes this clear. The spatial distance $d_t(P_1, P_2)$ between two particles P_1, P_2 at rest at time t is defined by using the spatial metric

$$ds_t^2 = \alpha(t)^2\big((1 - kr^2)^{-1}\,dr^2 + r^2(d\theta^2 + \sin^2\theta\,d\varphi^2)\big).$$

That is,

$$d_t(P_1, P_2) = \inf \int_{P_1}^{P_2} ds_t,$$

where the integrals are along paths from P_1 to P_2 at time t, and the infimum is taken over all such paths. We do not link this definition to any particular operational procedure for measuring distance. It is simply a geometric construction. By symmetry, if P_1 is at the origin, then the shortest path must be radial. So the distance in this case is[2]

$$d_t(P_1, P_2) = \alpha(t) \int_0^r \frac{dr'}{\sqrt{1 - kr'^2}},$$

where r is evaluated at P_2. Wherever P_1 and P_2 are located, we can conclude that one is inside the particle horizon of the other at time t if and only if

$$d_t(P_1, P_2) \leq \alpha(t) \int_{t_0}^t \frac{dt'}{\alpha(t')}.$$

Because of the symmetry between the two particles, we can also interpret the particle horizon the other way around: it separates particles that could have influenced events on the worldline at P_1 from those that could not. The particles beyond the particle horizon are beyond the knowledge of an observer at P_1 at time t.

For small values of r, the integral on the right in the definition (12.13) is approximately equal to r, whatever the value of k. In our cosmological models, α is given by (12.10) for small $t - t_0$, and therefore the radius of the particle horizon at time t, measured by the spatial metric at time t, is approximately

$$(t - t_0)^{2/(n+2)} \int_{t_0}^t \frac{dt'}{(t' - t_0)^{2/(n+2)}} = \frac{(n + 2)(t - t_0)}{n},$$

which goes to zero as $t \to t_0$.

Herein lies one of the puzzles of modern cosmology, the *horizon problem*. The cosmological models that emerge from the study of Robertson–Walker

---

[2] Note that in the case $k = 1$, the coordinate $r = \sin\chi$ has a maximum value $r = 1$, and therefore there is a maximum value for $d_t(P_1, P_2)$ of $\alpha(t)$. This is the distance between two antipodal particles on the three-sphere.

metrics have an initial singularity, the *big bang*, at which $\alpha \to 0$ and $\rho \to \infty$, provided at least that the energy condition in the singularity theorem holds. In these models, the universe was initially unimaginably hot, but it cooled as it expanded. At time t_r, a few hundred thousand years after the big bang, it was cool enough for electrons and nuclei to combine into atoms, a process called *recombination*. The black body radiation emitted at that time by the still extremely hot matter was free to travel through the universe thereafter, essentially without hindrance, and can be observed today, some 14 billion years after the big bang. As the universe expanded, the radiation was redshifted, and today it appears as the 'microwave background'. In every direction, we see black body radiation as if from a body at a temperature of 2.725°K. It is this observation, along with the isotropy of the radiation, that provides the most dramatic support for the big bang models. When we observe the radiation, we are literally looking at the hot matter filling the universe 14 billion years ago. It no longer looks so hot because of the cosmological redshift.

Let t denote the present time. Suppose radiation seen today at our galaxy P_0 was emitted by particle P_1 at time t_r. Then the current distance from P_0 to P_1 is (very nearly) the current size of the particle horizon. Therefore if radiation from the opposite direction was emitted by particle P_2 also at time t_r, then the current distance from P_1 to P_2 is approximately twice the current radius of the particle horizon. Therefore the distance from P_1 to P_2 at time t_r was approximately[3]

$$2\alpha(t_r) \int_{t_0}^{t} \frac{\mathrm{d}t'}{\alpha(t')} \, .$$

The problem is that under our assumptions about the equation of state this is very much greater than

$$\alpha(t_r) \int_{t_0}^{t_r} \frac{\mathrm{d}t'}{\alpha(t')} \, ,$$

which was the radius of the particle horizon at recombination. So no event at P_1 could have influenced P_2 by time t_r How then can the temperature of the radiation from the two directions be the same?

The problem arises from the behaviour of α as one approaches the big bang, and ultimately from the assumptions made about the equation of state. Near the big bang, however, temperatures and densities are unimaginably large, and conventional assumptions about the nature of matter are almost certainly inappropriate. The inflationary hypothesis hangs on the possibility that $k = 0$ and that in the early universe, the equation of state is rather different, and

---

[3] A possibility that we have brushed aside here, but which should be explored and eliminated, is that in a closed universe, P_1 can be close to P_2 even though both are at a great distance from P_0.

there is a time interval (t_1, t_2) before recombination in which the metric is the de Sitter metric (12.11). In this case

$$\alpha(t_r) \int_{t_0}^{t_r} \frac{dt'}{\alpha(t')} > \alpha(t_2) \int_{t_1}^{t_2} \frac{dt'}{\alpha(t')} = e^{H(t_2 - t_1)} - 1 \,.$$

Provided that the 'inflationary period' $t_2 - t_1$ during which the metric has this form is long enough, the radius of the particle horizon can be arbitrarily large at recombination. The horizon problem is then resolved.

Inflationary cosmology also addresses two other puzzles, the *smoothness problem* and the *flatness problem*, the observations that matter appears to be uniformly distributed and that the spatial geometry of the universe is very close to being flat. Both are unexpected other than in a universe evolving from finely tuned initial conditions. Guth [7] gives an account of the ideas; there is also an interesting critique in Penrose [18]. Perhaps the strongest lesson is that conditions in the early universe have observable consequences at the present time, and therefore observations on a cosmological scale—for example, of the fine details of the microwave background—can reveal information about the behaviour of matter under very extreme conditions, and therefore provide tests for ideas in particle physics.

Appendix A: Notes on Exercises

1.1   By spherical symmetry, the gravitational field $\boldsymbol{F}$ (the gravitational force per unit mass) must be of the form

$$\boldsymbol{F} = F(r)\hat{\boldsymbol{r}}\,,$$

where F is a function only of the distance r from the centre, and $\hat{\boldsymbol{r}}$ is the unit vector along the radius. Let S_r be the sphere of radius r with its centre at the centre of the body. By Gauss's theorem,

$$\int_{S_r} \boldsymbol{F}\,.\,\mathrm{d}\boldsymbol{S} \;=\; 4\pi r^2 F(r)$$

$$= \begin{cases} -4\pi Gm & r \geq a \\ -\dfrac{4\pi Gm r^3}{a^3} & r < a \end{cases}.$$

Hence

$$F(r) \;=\; \begin{cases} -\dfrac{Gm}{r^2} & r \geq a \\ -\dfrac{mr}{a^3} & r < a \end{cases}.$$

By solving $F = -\mathrm{d}\phi/\mathrm{d}r$, and by using the boundary conditions that ϕ should be continuous at $r = a$, and should go to zero at infinity, we deduce that

$$\phi = \begin{cases} -\dfrac{Gm}{r} & r \geq a \\ -\dfrac{Gmr(3a^2 - r^2)}{2a^3} & r < a \end{cases}.$$

1.2   We take the test particle to have unit mass. Then the equation of motion is

$$\ddot{\boldsymbol{r}} = -\frac{Gm\boldsymbol{r}}{r^3}.$$

By taking first the scalar product with $\dot{\boldsymbol{r}}$ and then the vector product with $\dot{\boldsymbol{r}}$, we obtain

$$\frac{\mathrm{d}}{\mathrm{d}t}\left(\tfrac{1}{2}\dot{\boldsymbol{r}}\cdot\dot{\boldsymbol{r}}\right) = -\frac{Gm\dot{r}}{r^2}, \qquad \frac{\mathrm{d}}{\mathrm{d}t}\left(\boldsymbol{r}\wedge\dot{\boldsymbol{r}}\right) = 0.$$

These integrate to give the energy and angular momentum equations. The second also implies that the motion is planar (either in a line, or in the plane orthogonal to the angular momentum vector $\boldsymbol{h} = \boldsymbol{r}\wedge\dot{\boldsymbol{r}}$). In plane polars, the two equations are

$$\tfrac{1}{2}(\dot{r}^2 + r^2\dot{\theta}^2) - \frac{Gm}{r} = E, \qquad r^2\dot{\theta} = J,$$

where E and J are constant.

With $u = Gm/r$, we have $p = -Gm\dot{r}/r^2\dot{\theta} = -Gm\dot{r}/J$. Hence the energy equation can be rewritten

$$\tfrac{1}{2}(p^2 + u^2) - \beta^2 u = k,$$

where $k = G^2 m^2 E/J^2$.

The constant J is the angular momentum, and the constant k is (a multiple of) the energy divided by the square of the angular momentum. It is really the sign of k that is significant: it is negative for elliptic and circular orbits, and positive for hyperbolic orbits; for parabolic orbits, it vanishes. In all cases, the curves in the p, u-plane are circles.

In the first case, $k > 0$. All the circles pass through the same two points on the p-axis. These are hyperbolic orbits. They reach infinity with nonzero kinetic energy (nonzero p). In the second case, $k = 0$ and the orbits just reach infinity ($u = 0$), but with no kinetic energy (because $p = 0$). The circles all touch the p-axis at the origin. These are the parabolic orbits. In the third case, $k < 0$ and the orbits are closed (none reaches $u = 0$; that is, $r = \infty$). The point $(u, p) = (\sqrt{-2k}, 0)$ corresponds to the circular orbit. The other circles correspond to elliptical orbits, with the intersection points with the u axis (i.e., the roots of $u^2 - 2\beta^2 u - 2k$) giving the perihelion and aphelion.

1.3   The actual gravitational field is $(0, -g)$. The acceleration of the upper end is $(f\cos\alpha, -f\sin\alpha)$. The apparent gravitational field in a frame moving with the upper end is the difference of these two vectors. That is,

$$(-f\cos\alpha, -g + f\sin\alpha).$$

The required condition is that the initial (vertical) and final (horizontal) positions of the pendulum should make the same angle with this vector; that is, the two components of this vector should have the same magnitude. (Think of a pendulum moving in the earth's gravitational field, with no acceleration, from an initial position making an angle β with the direction of gravity. It comes to rest in the opposite position making the same angle β with the direction of gravity.)

1.4   The behaviour is the same as in the absence of gravity: the ball stays where it is, relative to the bucket.

1.5   Hold the apparatus vertical, with the cup at the top, by the bottom of the tube. Then let the tube fall through your hand, grasping it again at the top. While it is falling, the apparent gravity vanishes, and the elastic string is able to draw the ball into the cup.

2.1   In (ii) and (vii), the free indices do not balance. In (iv), there are too many repetitions of c for the summation convention to be unambiguous. The others make sense.

2.3   First show that for any four 4-vectors T^a, X^a, Y^a, Z^a, we have

$$\varepsilon_{abcd}T^aX^bY^cZ^d = \begin{vmatrix} T^0 & X^0 & Y^0 & Z^0 \\ T^1 & X^1 & Y^1 & Z^1 \\ T^2 & X^2 & Y^2 & Z^2 \\ T^3 & X^3 & Y^3 & Z^3 \end{vmatrix}.$$

In principle, you do this by comparing the 24 nonzero terms of the sum on the left-hand side with the 24 terms of the expansion of the determinant on the right-hand side. But you can avoid this task by arguing that it is enough to consider special cases because both sides are multilinear in the four 4-vectors.

2.4   (i) Note that if $L^a{}_b$ is a (proper) Lorentz transformation, and if

$$\hat{T}^a = L^a{}_bT^b, \qquad \hat{X}^a = L^a{}_bX^b, \qquad \hat{Y}^a = L^a{}_bY^b, \qquad \hat{Z}^a = L^a{}_bZ^b,$$

then

$$\begin{vmatrix} \hat{T}^0 & \hat{X}^0 & \hat{Y}^0 & \hat{Z}^0 \\ \hat{T}^1 & \hat{X}^1 & \hat{Y}^1 & \hat{Z}^1 \\ \hat{T}^2 & \hat{X}^2 & \hat{Y}^2 & \hat{Z}^2 \\ \hat{T}^3 & \hat{X}^3 & \hat{Y}^3 & \hat{Z}^3 \end{vmatrix} = \det(L) \begin{vmatrix} T^0 & X^0 & Y^0 & Z^0 \\ T^1 & X^1 & Y^1 & Z^1 \\ T^2 & X^2 & Y^2 & Z^2 \\ T^3 & X^3 & Y^3 & Z^3 \end{vmatrix}.$$

Hence, because $\det(L) = 1$, we have

$$\begin{aligned} \varepsilon_{abcd}T^aX^bY^cZ^d &= \varepsilon_{abcd}\hat{T}^a\hat{X}^b\hat{Y}^c\hat{Z}^d \\ &= \varepsilon_{pqrs}L^p{}_aT^aL^q{}_bX^bL^r{}_cY^cL^s{}_dZ^d. \end{aligned}$$

Because this holds for any four 4-vectors, we have

$$\varepsilon_{abcd} = \varepsilon_{pqrs} L^p{}_a L^q{}_b L^r{}_c L^s{}_d \,,$$

which is the required tensor transformation law.

(ii) We have

$$\varepsilon^{abcd} = \begin{cases} -1 & \text{if } a,b,c,d \text{ is an even permutation of } 0,1,2,3 \\ 1 & \text{if } a,b,c,d \text{ is an odd permutation of } 0,1,2,3 \\ 0 & \text{otherwise} \end{cases} \,.$$

(iii) There are 4^4 terms in the sum $\varepsilon_{abcd}\varepsilon^{abcd}$, of which only 24 are nonzero (those with a,b,c,d a permutation of 0,1,2,3). All the nonzero terms are equal to -1, whether the permutation is even or odd. Hence the first identity.

2.6   We have

$$\left(F_{ab}\right) = \begin{pmatrix} 0 & E_1 & E_2 & E_3 \\ -E_1 & 0 & -B_3 & B_2 \\ -E_2 & B_3 & 0 & -B_1 \\ -E_3 & -B_2 & B_1 & 0 \end{pmatrix}$$

$$\left(F^{ab}\right) = \begin{pmatrix} 0 & -E_1 & -E_2 & -E_3 \\ E_1 & 0 & -B_3 & B_2 \\ E_2 & B_3 & 0 & -B_1 \\ E_3 & -B_2 & B_1 & 0 \end{pmatrix}$$

$$\left(F^{*ab}\right) = \begin{pmatrix} 0 & B_1 & B_2 & B_3 \\ -B_1 & 0 & -E_3 & E_2 \\ -B_2 & E_3 & 0 & -E_1 \\ -B_3 & -E_2 & E_1 & 0 \end{pmatrix} \,.$$

The scalar $F_{ab}F^{ab}$ is the sum of the products of the entries in the first matrix with the corresponding entries in the second; that is, $2(\boldsymbol{B}.\boldsymbol{B} - \boldsymbol{E}.\boldsymbol{E})$.

Similarly, $F_{ab}F^{*ab}$ is the sum of the products of the entries in the first matrix with those in the third; that is $4\boldsymbol{E}.\boldsymbol{B}$.

Because both scalars are invariants, it follows that $\boldsymbol{B}.\boldsymbol{B} - \boldsymbol{E}.\boldsymbol{E}$ and $\boldsymbol{E}.\boldsymbol{B}$ are invariants.

2.7   In the observer's rest frame,

$$\left(U^a\right) = \begin{pmatrix} 1 \\ 0 \\ 0 \\ 0 \end{pmatrix} \qquad \left(F^*_{ab}\right) = \begin{pmatrix} 0 & -B_1 & -B_2 & -B_3 \\ B_1 & 0 & -E_3 & E_2 \\ B_2 & E_3 & 0 & -E_1 \\ B_3 & -E_2 & E_1 & 0 \end{pmatrix} ,$$

where $\boldsymbol{E}$ and $\boldsymbol{B}$ are the electric and magnetic fields seen by the observer. Now

$$\left(F_{ab}^* U^b\right) = \begin{pmatrix} 0 & -B_1 & -B_2 & -B_3 \\ B_1 & 0 & -E_3 & E_2 \\ B_2 & E_3 & 0 & -E_1 \\ B_3 & -E_2 & E_1 & 0 \end{pmatrix} \begin{pmatrix} 1 \\ 0 \\ 0 \\ 0 \end{pmatrix} = \begin{pmatrix} 0 \\ B_1 \\ B_2 \\ B_3 \end{pmatrix}.$$

Therefore $F_{ab}^* U^b = 0$ if and only if the observed magnetic field vanishes.

In a general frame

$$\left(U^a\right) = \gamma(u) \begin{pmatrix} 1 \\ u_1 \\ u_2 \\ u_3 \end{pmatrix} \qquad \left(F_{ab}^*\right) = \begin{pmatrix} 0 & -B_1 & -B_2 & -B_3 \\ B_1 & 0 & -E_3 & E_2 \\ B_2 & E_3 & 0 & -E_1 \\ B_3 & -E_2 & E_1 & 0 \end{pmatrix},$$

where $\boldsymbol{u}$ is the observer's velocity relative to the frame and $\boldsymbol{E}$ and $\boldsymbol{B}$ are now the electric and magnetic fields in the frame. In this frame, we have

$$\left(F_{ab}^* U^b\right) = \gamma(u) \begin{pmatrix} -\boldsymbol{B}.\boldsymbol{u} \\ B_1 - u_2 E_3 + u_3 E_2 \\ B_2 - u_3 E_1 + u_1 E_3 \\ B_3 - u_1 E_2 + u_2 E_1 \end{pmatrix}.$$

Thus the observer sees zero magnetic field if and only if

$$\boldsymbol{B}.\boldsymbol{u} = 0, \qquad \boldsymbol{B} - \boldsymbol{u} \wedge \boldsymbol{E} = 0.$$

To determine whether there exists a frame in which the observed magnetic field vanishes, we have to determine whether these equations can be solved for $\boldsymbol{u}$ with $\boldsymbol{E}$ and $\boldsymbol{B}$ given.

Clearly there is no solution unless $\boldsymbol{E}.\boldsymbol{B} = 0$. When this condition holds, the second equation implies

$$\boldsymbol{u} = \frac{\boldsymbol{E} \wedge \boldsymbol{B}}{\boldsymbol{E}.\boldsymbol{E}} + \lambda \boldsymbol{E}$$

for some $\lambda \in \mathbb{R}$. The two equations are satisfied for any choice of λ; however, $|\boldsymbol{u}|$ is minimal when $\lambda = 0$. In this case

$$|\boldsymbol{u}| = \left| \frac{\boldsymbol{E} \wedge \boldsymbol{B}}{\boldsymbol{E}.\boldsymbol{E}} \right| = \frac{|\boldsymbol{B}|}{|\boldsymbol{E}|}.$$

Hence there is a solution with $|\boldsymbol{u}| < 1$ if and only if $\boldsymbol{E}.\boldsymbol{B} = 0$ and $\boldsymbol{B}.\boldsymbol{B} < \boldsymbol{E}.\boldsymbol{E}$.

3.1 Show first that if there are two rest-velocities, then the corresponding rest densities must be equal, because otherwise the rest-velocities would be orthogonal, which is not possible for timelike vectors. By taking a linear combination, deduce that there is a null four-vector N^a such that $T^{ab}N_aN_b = 0$. Now obtain a contradiction with $\rho_V \to \infty$ as $v \to 1$.

3.3 Why is it enough to consider only τ^{00}?

4.1 By the chain rule,
$$\frac{\partial \tilde{x}^a}{\partial x^p}\frac{\partial x^p}{\partial \tilde{x}^b} = \delta^a_b\,.$$
Therefore, by differentiating with respect to $\tilde{x}^c$,
$$\frac{\partial \tilde{x}^a}{\partial x^p}\frac{\partial^2 x^p}{\partial \tilde{x}^b \partial \tilde{x}^c} + \frac{\partial x^q}{\partial \tilde{x}^c}\frac{\partial}{\partial x^q}\left(\frac{\partial \tilde{x}^a}{\partial x^p}\right)\frac{\partial x^p}{\partial \tilde{x}^b} = 0\,.$$
The result follows.

4.2 We have to show that the components transform correctly.
$$\begin{aligned}
&X^b\partial_b Y^a - Y^b\partial_b X^a\\
&= \tilde{X}^c\frac{\partial x^b}{\partial \tilde{x}^c}\frac{\partial}{\partial x^b}\left(\frac{\partial x^a}{\partial \tilde{x}^d}\tilde{Y}^d\right) - \tilde{Y}^c\frac{\partial x^b}{\partial \tilde{x}^c}\frac{\partial}{\partial x^b}\left(\frac{\partial x^a}{\partial \tilde{x}^d}\tilde{X}^d\right)\\
&= \tilde{X}^c\frac{\partial}{\partial \tilde{x}^c}\left(\frac{\partial x^a}{\partial \tilde{x}^d}\tilde{Y}^d\right) - \tilde{Y}^c\frac{\partial}{\partial \tilde{x}^c}\left(\frac{\partial x^a}{\partial \tilde{x}^d}\tilde{X}^d\right)\\
&= \left(\tilde{X}^c\tilde{\partial}_c\tilde{Y}^d - \tilde{Y}^c\tilde{\partial}_c\tilde{X}^d\right)\frac{\partial x^a}{\partial \tilde{x}^d} - \left(\tilde{X}^c\tilde{Y}^d - \tilde{X}^d\tilde{Y}^c\right)\frac{\partial^2 x^a}{\partial \tilde{x}^c \partial \tilde{x}^d}\\
&= \left(\tilde{X}^c\tilde{\partial}_c\tilde{Y}^d - \tilde{Y}^c\tilde{\partial}_c\tilde{X}^d\right)\frac{\partial x^a}{\partial \tilde{x}^d}\,,
\end{aligned}$$
because the last partial derivative in the penultimate line is symmetric in c, d.

The vector field Z is called the *Lie bracket* of X and Y, and is usually denoted by $[X, Y]$.

4.3 The Lagrange equations are
$$\frac{\mathrm{d}V^a}{\mathrm{d}\tau} + \Gamma^a_{bc}V^bV^c = 0\,,$$
where $V^a = \mathrm{d}x^a/\mathrm{d}\tau$. Hence
$$\frac{\mathrm{d}}{\mathrm{d}\tau}\left(g_{ab}V^aV^b\right) = -2\Gamma^a_{bc}V^bV^cV_a + V^aV^b\left(\partial_c g_{ab}\right)V^c = 0\,,$$
because
$$\Gamma^a_{bc}V^bV^cV_a = \tfrac{1}{2}V_aV^bV^cg^{ad}\left(\partial_b g_{dc} + \partial_c g_{db} - \partial_d g_{bc}\right) = \tfrac{1}{2}V^aV^bV^c\left(\partial_c g_{ab}\right)\,.$$

Alternatively, the result follows from the fact that L has no explicit dependence on proper time (i.e. $\partial L/\partial \tau = 0$). Hence the corresponding Hamiltonian is conserved. But because the Lagrangian is a homogeneous quadratic in the $\dot{x}^a$s, the Hamiltonian and the Lagrangian coincide.

4.4   In this metric, the Lagrangian for the geodesics is

$$L = \tfrac{1}{2}\left(\dot{t}^2 - \dot{r}^2 - \sin^2 r\,\dot{\theta}^2 - \sin^2 r\,\sin^2\theta\,\dot{\phi}^2\right).$$

So the geodesic equations are

$$\ddot{t} = 0$$
$$\ddot{r} - \sin r\cos r\,\dot{\theta}^2 - \sin r\cos r\sin^2\theta\,\dot{\phi}^2 = 0$$
$$\ddot{\theta} - \sin\theta\cos\theta\,\dot{\phi}^2 + 2\cot r\,\dot{\theta}\dot{r} = 0$$
$$\ddot{\phi} + 2\cot r\,\dot{\phi}\dot{r} + 2\cot\theta\,\dot{\theta}\dot{\phi} = 0.$$

We read off from these that the nonzero Christoffel symbols are:

$$\Gamma^1_{22} = -\sin r\cos r, \quad \Gamma^1_{33} = -\sin r\cos r\sin^2\theta$$
$$\Gamma^2_{12} = \Gamma^2_{21} = \cot r, \quad \Gamma^2_{33} = -\sin\theta\cos\theta$$
$$\Gamma^3_{23} = \Gamma^3_{32} = \cot\theta, \quad \Gamma^3_{13} = \Gamma^3_{31} = \cot r.$$

The geodesic equations are consistent with $r = \theta = \pi/2$ because $\cos(\pi/2) = 0$. With this condition, they reduce to $\ddot{\phi} = 0 = \ddot{t}$.

The model is incorrect because the universe is expanding; it required an awkward modification of the gravitational field equations, which he later described as the biggest mistake of his life.

4.5   The transformation to T, W, X, Y, Z from the *hyperspherical* coordinates t, r, θ, ϕ gives

$$\mathrm{d}T = \mathrm{d}t$$
$$\mathrm{d}W = -\sin r\,\mathrm{d}r$$
$$\mathrm{d}X = \cos r\sin\theta\sin\phi\,\mathrm{d}r + \sin r\cos\theta\sin\phi\,\mathrm{d}\theta + \sin r\sin\theta\cos\phi\,\mathrm{d}\phi$$
$$\mathrm{d}Y = \cos r\sin\theta\cos\phi\,\mathrm{d}r + \sin r\cos\theta\cos\phi\,\mathrm{d}\theta - \sin r\sin\theta\sin\phi\,\mathrm{d}\phi$$
$$\mathrm{d}Z = \cos r\cos\theta\,\mathrm{d}r - \sin r\sin\theta\,\mathrm{d}\theta,$$

from which one gets $\mathrm{d}S^2 = \mathrm{d}s^2$. The transformation maps the Einstein universe into the product of the T-axis and the three-sphere

$$S^3 = \{W^2 + X^2 + Y^2 + Z^2 = 1\}.$$

The tasks 'What portion ...' and 'Deduce that ...' are not precisely defined as they stand, because neither the manifold on which the Einstein metric is defined nor the ranges of the coordinates have been specified. They are intended to provoke consideration of the analogous coordinate systems on the surface of the earth. The hyperspherical coordinates (and t) define a chart on almost all of the product of the T-axis and the three-sphere, less, for example $X = 0$, $Y \leq 0$. This corresponds to making the natural choices in which the manifold is $\mathbb{R} \times S^3$ and the ranges of the coordinates are

$$-\infty < t < \infty, \quad 0 < r < \pi, \quad 0 < \theta < \pi, \quad -\pi < \phi < \pi.$$

The geodesics on which $r = \theta = \pi/2$ are mapped to

$$X^2 + Y^2 = 1.$$

Because $\ddot{t} = 0$, they are the paths given by travelling at constant speed on a great circle on the three-sphere. By rotational symmetry, all geodesics are of this form.

This is very similar to the relationship between the sphere metric $d\theta^2 + \sin^2\theta \, d\varphi^2$ and the Euclidean metric $dx^2 + dy^2 + dz^2$, given by $x = \sin\theta\cos\varphi$, $y = \sin\theta\sin\varphi$, $z = \cos\theta$. Thus the Einstein universe is a curved 'hypersurface' in a 'flat' five-dimensional space, in the same way that the sphere is a curved surface in three-dimensional Euclidean space.

5.2 Since U, V, W are coplanar, it is only necessary to check that the inner products of both sides with U and V are equal. For the inner product with U, we use

$$
\begin{aligned}
g(U, V\sinh\theta_{OQ} - U\sinh\theta_{PQ}) &= \cosh\theta_{OP}\sinh\theta_{OQ} - \sinh\theta_{PQ} \\
&= \cosh\theta_{OP}\sinh\theta_{OQ} - \sinh(\theta_{OQ} - \theta_{OP}) \\
&= \cosh\theta_{OQ}\sinh\theta_{OP},
\end{aligned}
$$

by using the identity $\sinh(A - B) = \sinh A \cosh B - \sinh B \cosh A$.

5.3 In local inertial coordinates at an event,

$$X^b\nabla_b Y^a - Y^b\nabla_b X^a = X^b\partial_b Y^a - Y^b\partial_b X^a.$$

5.4 We have

$$
\begin{aligned}
\partial_a \alpha_b &= \partial_a\left(\frac{\partial\tilde{x}^d}{\partial x^b}\tilde{\alpha}_d\right) \\
&= \frac{\partial^2\tilde{x}^d}{\partial x^a\partial x^b}\tilde{\alpha}_d + \frac{\partial\tilde{x}^d}{\partial x^b}\partial_a\tilde{\alpha}_d \\
&= \frac{\partial^2\tilde{x}^d}{\partial x^a\partial x^b}\tilde{\alpha}_d + \frac{\partial\tilde{x}^c}{\partial x^a}\frac{\partial\tilde{x}^d}{\partial x^b}\tilde{\partial}_c\tilde{\alpha}_d.
\end{aligned}
$$

The result follows because the second partial derivative is symmetric in a, b. The other part follows from

$$\nabla_a \alpha_b - \nabla_b \alpha_a = \partial_a \alpha_b - \Gamma_{ab}^c \alpha_c - \partial_b \alpha_a + \Gamma_{ba}^c \alpha_c = \partial_a \alpha_b - \partial_b \alpha_a$$

because $\Gamma_{ab}^c = \Gamma_{ba}^c$.

5.5   We have

$$
\begin{aligned}
\nabla_a g_{bc} &= \partial_a g_{bc} - g_{dc} \Gamma_{ab}^d - g_{bd} \Gamma_{ac}^d \\
&= \partial_a g_{bc} - K_{cab} - K_{bac}.
\end{aligned}
$$

We note that $K_{cab} = K_{cba}$ because ∇ is torsion-free. Therefore

$$
\begin{aligned}
\partial_a g_{bc} &= K_{bac} + K_{cab} \\
\partial_b g_{ca} &= K_{cba} + K_{abc} \\
\partial_c g_{ab} &= K_{acb} + K_{bca}.
\end{aligned}
$$

By adding the last two equations and subtracting the first, we obtain

$$2 K_{abc} = \partial_b g_{ca} + \partial_c g_{ab} - \partial_a g_{bc},$$

from which it follows that Γ_{bc}^a is the Christoffel symbol.

5.6   If we substitute

$$g_{ab} = \frac{\partial \tilde{x}^c}{\partial x^a} \frac{\partial \tilde{x}^d}{\partial x^b} \tilde{g}_{cd}, \qquad g^{ab} = \frac{\partial x^a}{\partial \tilde{x}^c} \frac{\partial x^b}{\partial \tilde{x}^d} \tilde{g}^{cd}$$

into

$$\Gamma_{bc}^a = \tfrac{1}{2} g^{ad} \Big(\partial_b g_{cd} + \partial_c g_{bd} - \partial_d g_{bc} \Big),$$

then we get

$$
\begin{aligned}
\Gamma_{bc}^a &= \tilde{\Gamma}_{ef}^d \frac{\partial x^a}{\partial \tilde{x}^d} \frac{\partial \tilde{x}^e}{\partial x^b} \frac{\partial \tilde{x}^f}{\partial x^c} + \tilde{g}^{pq} \tilde{g}_{rs} \frac{\partial x^a}{\partial \tilde{x}^p} \frac{\partial x^d}{\partial \tilde{x}^q} \left[\frac{\partial}{\partial x^b} \left(\frac{\partial \tilde{x}^r}{\partial x^c} \frac{\partial \tilde{x}^s}{\partial x^d} \right) \right. \\
&\qquad \left. + \frac{\partial}{\partial x^c} \left(\frac{\partial \tilde{x}^r}{\partial x^b} \frac{\partial \tilde{x}^s}{\partial x^d} \right) - \frac{\partial}{\partial x^d} \left(\frac{\partial \tilde{x}^r}{\partial x^b} \frac{\partial \tilde{x}^s}{\partial x^c} \right) \right] \\
&= \tilde{\Gamma}_{ef}^d \frac{\partial x^a}{\partial \tilde{x}^d} \frac{\partial \tilde{x}^e}{\partial x^b} \frac{\partial \tilde{x}^f}{\partial x^c} + \frac{\partial x^a}{\partial \tilde{x}^p} \frac{\partial^2 \tilde{x}^p}{\partial x^b \partial x^c} + \tfrac{1}{2} \tilde{g}^{pq} \tilde{g}_{rs} \frac{\partial x^a}{\partial \tilde{x}^p} \frac{\partial x^d}{\partial \tilde{x}^q} \left(\frac{\partial \tilde{x}^r}{\partial x^c} \frac{\partial^2 \tilde{x}^s}{\partial x^b \partial x^d} \right. \\
&\qquad \left. + \frac{\partial \tilde{x}^r}{\partial x^b} \frac{\partial^2 \tilde{x}^s}{\partial x^c \partial x^d} - \frac{\partial \tilde{x}^r}{\partial x^b} \frac{\partial^2 \tilde{x}^s}{\partial x^d \partial x^c} - \frac{\partial \tilde{x}^s}{\partial x^c} \frac{\partial^2 \tilde{x}^r}{\partial x^d \partial x^b} \right) \\
&= \tilde{\Gamma}_{ef}^d \frac{\partial x^a}{\partial \tilde{x}^d} \frac{\partial \tilde{x}^e}{\partial x^b} \frac{\partial \tilde{x}^f}{\partial x^c} + \frac{\partial x^a}{\partial \tilde{x}^p} \frac{\partial^2 \tilde{x}^p}{\partial x^b \partial x^c}.
\end{aligned}
$$

5.7   For any square matrix B, we have

$$\det(1 + \epsilon B) = 1 + \epsilon \operatorname{tr}(B) + O(\epsilon^2),$$

where 1 denotes the identity matrix. Therefore

$$\begin{aligned}
\det(A + \epsilon B) &= \det\Big(A\big(1 + \epsilon A^{-1}B\big)\Big) + O(\epsilon^2) \\
&= \det A\Big(1 + \epsilon \operatorname{tr}\big(A^{-1}B\big)\Big) + O(\epsilon^2).
\end{aligned}$$

Therefore if A is a function of t,

$$\frac{1}{\det A}\frac{\mathrm{d}}{\mathrm{d}t}\big(\det A\big) = \operatorname{tr}\left(A^{-1}\frac{\mathrm{d}A}{\mathrm{d}t}\right).$$

We conclude that

$$\partial_a \log|g| = g^{bc}\partial_a g_{bc},$$

because the right-hand side is the trace of $A^{-1}\partial_a A$ when A is the matrix with entries g_{bc}. Therefore

$$\Gamma^b_{ab} = \tfrac{1}{2}g^{bd}\big(\partial_b g_{ad} + \partial_a g_{bd} - \partial_d g_{ab}\big) = \tfrac{1}{2}g^{bd}\partial_a g_{bd} = \partial_a \log\sqrt{|g|}.$$

In general coordinates on flat space–time

$$\begin{aligned}
\nabla_a \nabla^a u &= \nabla_a\Big(g^{ab}\partial_b u\Big) \\
&= \partial_a\Big(g^{ab}\partial_b u\Big) + \Gamma^a_{ac}g^{cb}\partial_b u \\
&= \partial_a\Big(g^{ab}\partial_b u\Big) + \partial_a\Big(\log\sqrt{|g|}\Big)g^{ab}\partial_b u \\
&= |g|^{-1/2}\partial_a\Big(|g|^{1/2}g^{ab}\partial_b u\Big).
\end{aligned}$$

The left-hand side is invariant. In inertial coordinates, the first line gives

$$\nabla_a \nabla^a u = \frac{\partial^2 u}{\partial t^2} - \frac{\partial^2 u}{\partial x^2} - \frac{\partial^2 u}{\partial y^2} - \frac{\partial^2 u}{\partial z^2},$$

which is the wave equation.

In spherical polars, we have

$$\mathrm{d}s^2 = \mathrm{d}t^2 - \mathrm{d}r^2 - r^2\,\mathrm{d}\theta^2 - r^2\sin^2\theta\,\mathrm{d}\phi^2,$$

from which we have $|g|^{1/2} = r^2\sin\theta$. We also have

$$\big(g^{ab}\big) = \begin{pmatrix} 1 & & & \\ & -1 & & \\ & & -1/r^2 & \\ & & & -1/r^2\sin^2\theta \end{pmatrix}.$$

Therefore the wave equation is

$$\frac{\partial^2 u}{\partial t^2} - \frac{1}{r^2}\frac{\partial}{\partial r}\left(r^2 \frac{\partial u}{\partial r}\right) - \frac{1}{r^2 \sin\theta}\frac{\partial}{\partial\theta}\left(\sin\theta\,\frac{\partial u}{\partial\theta}\right) - \frac{1}{r^2 \sin^2\theta}\frac{\partial^2 u}{\partial\phi^2} = 0\,.$$

5.8 Suppose that there exists such a coordinate system. Then

$$\begin{aligned}\nabla_a X^d &= \partial_a X^d + \tfrac{1}{2}X^c g^{bd}\left(\partial_a g_{bc} + \partial_c g_{ba} - \partial_b g_{ca}\right)\\ &= \tfrac{1}{2}g^{bd}\left(\partial_a g_{0b} + \partial_0 g_{ba} - \partial_d g_{0b}\right).\end{aligned}$$

Hence

$$\nabla_a X_b = g_{bd}\nabla_a X^d = \tfrac{1}{2}\left(\partial_a g_{0b} - \partial_b g_{0a}\right),$$

and therefore $\nabla_a X_b + \nabla_b X_a = 0$.

Vector fields with this property are called *Killing vectors*: they arise from symmetries of space–time.

The corresponding result is $\nabla_a X_b + \nabla_b X_a = f g_{ab}$. By contracting with g^{ab}, we have $f = \tfrac{1}{2}\nabla_a X^a$ because $g_{ab}g^{ab} = 4$.

5.9 A skew-symmetric tensor changes sign when two indices are interchanged. Therefore any components with equal values of two indices must vanish. For a tensor with five indices, at least two must be equal. Therefore $T_{[bcde]} = 0$.

5.10 The 'number of independent components' is the dimension of the corresponding vector space of tensors.

(a) $\tfrac{1}{2}n(n-1)$, by the same argument as in the question, noting that the components with $a = b$ vanish by skew-symmetry.

(b) For $k \le n$, there are

$$\binom{n}{k}$$

independent ways of choosing k distinct values for the indices, and therefore that number of independent components. For $k > n$, the answer is zero, by the same argument as in Exercise 5.9.

(c) There are $\tfrac{1}{2}n(n-1)$ ways of choosing the first pair of indices and the same number of ways of choosing the second, so there are $\tfrac{1}{4}n^2(n-1)^2$ independent components.

(d) If we add $R_{abcd} = R_{cdab}$ to the conditions in (c), then the number is reduced to $\tfrac{1}{2}N(N+1)$, where $N = \tfrac{1}{2}n(n-1)$ (because this is the number of independent entries in an $N \times N$ symmetric matrix).

5.12 By definition,
$$\nabla_a \nabla_b X^c - \nabla_b \nabla_a X^c = R_{abd}{}^c X^d \,.$$

The result follows by lowering the index c, and by using $R_{abcd} = -R_{abdc}$.

$$
\begin{aligned}
\nabla_a \nabla_b X_c - \nabla_b \nabla_a X_c &= -R_{abcd} X^d \\
\nabla_b \nabla_c X_a - \nabla_c \nabla_b X_a &= -R_{bcad} X^d \\
\nabla_c \nabla_a X_b - \nabla_a \nabla_c X_b &= -R_{cabd} X^d
\end{aligned}
$$

(the second two equations are obtained from the first by cyclic permutation of a, b, c). By adding the first and last, and by subtracting the second,
$$2\nabla_a \nabla_b X_c = \left(-R_{abcd} + R_{bcad} - R_{cabd}\right) X^d \,.$$

But, by the symmetries of the Riemann tensor,
$$R_{abcd} + R_{bcad} + R_{cabd} = 0 \,.$$

By adding this to the expression in brackets above, we get
$$2\nabla_a \nabla_b X_c = 2R_{bcad} X^d \,.$$

By contracting the identity with $V^a V^b$, where $V^a = \mathrm{d}x^a / \mathrm{d}\tau$, we obtain
$$D^2 X_c = V^a V^b \nabla_a \nabla_b X_c = V^a V^b R_{bcad} X^d = R_{abdc} V^a X^b V^d \,,$$

which is the equation of geodesic deviation.

5.13 By Exercise 5.4, we have $F_{ab} = 2\partial_{[a}\Phi_{b]}$ everywhere, in any coordinate system. Thus
$$\nabla_{[a} F_{bc]} = 2\nabla_{[a}\partial_b \Phi_{c]} \,.$$

However, in local inertial coordinates at an event, this reduces to
$$2\partial_{[a}\partial_b \Phi_{c]}$$

at the event, which vanishes because partial derivatives commute. (In the language of differential forms, we have shown that $\mathrm{d}^2 = 0$.)

For the other Maxwell equation, we have
$$\nabla^a\left(\nabla_a \Phi_b - \nabla_b \Phi_a\right) = \Box \Phi_b + R^a{}_{bac}\Phi^c - \nabla_b\left(\nabla^a \Phi_a\right) \,.$$

Therefore the second set of Maxwell equations $\nabla^a F_{ab} = 0$ reduces to
$$\Box \Phi_b - \nabla_b\left(\nabla^a \Phi_a\right) = -R_{ab}\Phi^a \,.$$

5.14 Because $X_a = \nabla_a f$, we have $\nabla_a X_b = \nabla_b X_a$. Therefore,

$$X^a \nabla_a X^b = X^a \nabla^b X_a = \tfrac{1}{2} \nabla^b (X_a X^a) = 0 \,,$$

and so the curves are geodesics.

5.15 The geodesic equations are

$$\frac{\mathrm{d}}{\mathrm{d}\tau} \left(\tfrac{1}{2}\dot{v} + \log(x^2 + y^2)\dot{u} \right) = 0 \,,$$

together with

$$\ddot{u} = 0, \qquad \ddot{x} + \frac{x\dot{u}^2}{x^2 + y^2} = 0, \qquad \ddot{y} + \frac{y\dot{u}^2}{x^2 + y^2} = 0 \,.$$

The x and y equations are the same as those obtained from the Lagrangian

$$L = \tfrac{1}{2} \left(\dot{x}^2 + \dot{y}^2 - A^2 \log(x^2 + y^2) \right) = \tfrac{1}{2} \left(\dot{r}^2 + r^2 \dot{\theta}^2 - A^2 \log r^2 \right)$$

in classical mechanics, where A is the constant value of $\dot{u}$ and r, θ are plane polar coordinates. This is the Lagrangian of a central force problem with potential $V = A^2 \log r$. We have $K = x\dot{y} - y\dot{x} = r^2 \dot{\theta}$ is constant because $\partial L / \partial \theta = 0$. Also energy is conserved (because L has no explicit time dependence). Therefore

$$\tfrac{1}{2} \left(\dot{r}^2 + \frac{K^2}{r^2} \right) + A^2 \log r = \text{constant.}$$

However, for $K \neq 0$, we have $A^2 \log r + K^2/r^2 \to \infty$ as $r \to 0$, so no solution can reach $r = 0$.

7.1 The Schwarzschild metric is

$$\mathrm{d}s^2 = \left(1 - \frac{2m}{r} \right) \mathrm{d}t^2 - \frac{\mathrm{d}r^2}{1 - 2m/r} - r^2 \, \mathrm{d}\theta^2 - r^2 \sin^2 \theta \, \mathrm{d}\phi^2 \,.$$

The four-velocity of an observer at rest has components

$$(u^a) = (\dot{t}, 0, 0, 0) \,,$$

where

$$\dot{t} = \frac{\mathrm{d}t}{\mathrm{d}\tau} = \frac{1}{\sqrt{1 - 2m/r}} \,,$$

because $1 = g_{ab} U^a U^b = (1 - 2m/r)\dot{t}^2$.

Along a radial null geodesic, $\mathrm{d}s^2 = 0$, and θ and ϕ are constant. Therefore

$$\left(1 - \frac{2m}{r} \right) \mathrm{d}t^2 - \frac{\mathrm{d}r^2}{1 - 2m/r} = 0 \,,$$

and hence

$$\frac{dt}{dr} = \frac{r}{r - 2m}.$$

By integrating this, we find that the coordinate time t_1 at which the photon leaves C_1 is related to the coordinate time t_2 at which the photon arrives at C_2 by

$$t_2 - t_1 = \int_{r_1}^{r_2} \frac{r \, dr}{r - 2m}.$$

Because the right-hand side is independent of t_1, we have that the coordinate time interval Δt_1 between A and A' is the same as the coordinate time interval Δt_2 between B and B'. (This is essentially the same as the argument in the first chapter that gravitational redshift is incompatible with special relativity.) Therefore, by the formula above for $dt/d\tau$, the corresponding proper time intervals are related by

$$\frac{\Delta\tau_1}{\Delta\tau_2} = \frac{\sqrt{1 - 2m/r_1}}{\sqrt{1 - 2m/r_2}}.$$

When m is small the right-hand side is

$$1 - \frac{m}{r_1} + \frac{m}{r_2} = 1 - \frac{mh}{r_1^2},$$

where $h = r_2 - r_1$ and second-order terms in m/r and h/r are neglected. In SI units, this gives

$$\Delta\tau_1 = \Delta\tau_2 \left(1 - \frac{Gmh}{r_2^2 c^2}\right) = \Delta\tau_2 \left(1 - \frac{gh}{c^2}\right),$$

where $g = Gm/r_2^2$ is the acceleration due to gravity. By taking the approximate values (in SI units) $g = 10$, $h = 1$ (that is, 1m), $c = 3 \times 10^8$, and $\Delta\tau_2 = 3 \times 10^7$ (that is, one year in seconds), we get

$$\Delta\tau_1 \sim \Delta s_2 - 3 \times 10^{-9} \, \text{s}.$$

That is, the watch on your ankle (at $r = r_1$) appears to lose 3×10^{-9} seconds relative to the watch on your wrist (at $r = r_2 = r_1 + h$).

7.2   If ∇ is the Levi-Civita connection, then

$$\begin{aligned}
X^a \nabla_a T_{bc} &+ T_{ac} \nabla_b X^a + T_{ba} \nabla_c X^a \\
&= X^a \partial_a T_{bc} + T_{ac} \partial_b X^a + T_{ba} \partial_c X^a \\
&\quad + X^a \Gamma_{ab}^d T_{dc} - T_{ac} \Gamma_{bd}^a X^d - T_{ba} \Gamma_{cd}^a X^d + X^a \Gamma_{ac}^d T_{bd} \\
&= X^a \partial_a T_{bc} + T_{ac} \partial_b X^a + T_{ba} \partial_c X^a,
\end{aligned}$$

because $\Gamma_{ab}^d = \Gamma_{ba}^d$. Hence the result follows from the fact that the left-hand side is a tensor.

We know that if $Z = [X, Y]$, then

$$Z^a = X^b \nabla_b Y^a - Y^b \nabla_b X^a .$$

Hence

$$\begin{aligned}
\nabla_a & Z_b + \nabla_b Z_a \\
&= \nabla_a \left(X^c \nabla_c Y_b - Y^c \nabla_c X_b \right) + \nabla_b \left(X^c \nabla_c Y_a - Y^c \nabla_c X_a \right) \\
&= \nabla_a X^c \nabla_c Y_b - \nabla_a Y^c \nabla_c X_b + \nabla_b X^c \nabla_c Y_a - \nabla_b Y^c \nabla_c X_a \\
&\quad + X^c \nabla_a \nabla_c Y_b - Y^c \nabla_a \nabla_c X_b + X^c \nabla_b \nabla_c Y_a - Y^c \nabla_b \nabla_c X_a \\
&= -R_{cbad} X^c Y^d + R_{cbad} X^d Y^c - R_{cabd} X^c Y^d + R_{cabd} Y^c X^d \\
&= 0 ,
\end{aligned}$$

by Exercise 5.12, together with the symmetries of the Riemann tensor.

Another method is to prove first that

$$\mathcal{L}_X \mathcal{L}_y - \mathcal{L}_Y \mathcal{L}_X = \mathcal{L}_{[X,Y]} ,$$

where the operator $\mathcal{L}_X$ (the 'Lie derivative') is defined on tensors of type (0,2) by

$$\mathcal{L}_X T_{bc} = X^a \partial_a T_{bc} + T_{ac} \partial_b X^a + T_{ba} \partial_c X^a .$$

We now have the fact that $C = X_a \dot{x}^a$ is constant along a geodesic $x^a(s)$ if and only if X is a Killing vector. Now

If X has components $(1, 0, 0, 0)$ then $C = (1 - 2m/r) \dot{t}$;
If X has components $(0, 0, 0, 1)$ then $C = -r^2 \sin^2 \theta \, \dot{\phi}$;
If X has components $(0, 0 - \cos\phi, \cot\theta \sin\phi)$ then

$$C = r^2 \cos\phi \, \dot{\theta} - r^2 \sin^2 \theta \cot\theta \sin\phi \, \dot{\phi} .$$

The first two are clearly constant because the geodesic Lagrangian

$$L = \tfrac{1}{2} \left[\left(1 - \frac{2m}{r} \right) \dot{t}^2 - \frac{\dot{r}^2}{1 - 2m/r} - r^2 \dot{\theta}^2 - r^2 \sin^2 \theta \, \dot{\phi}^2 \right]$$

is independent of t and ϕ. In the third case, we use the two geodesic equations

$$\frac{\mathrm{d}}{\mathrm{d}\tau} \left(-r^2 \sin^2 \theta \, \dot{\phi} \right) = 0 \qquad \frac{\mathrm{d}}{\mathrm{d}\tau} \left(-r^2 \dot{\theta} \right) + r^2 \sin\theta \cos\theta \, \dot{\phi}^2 = 0$$

to deduce that

$$\frac{\mathrm{d}C}{\mathrm{d}\tau} = \frac{\mathrm{d}}{\mathrm{d}\tau}\left(r^2\dot\theta\right)\cos\phi - r^2\sin\phi\,\dot\theta\dot\phi$$
$$- r^2\sin^2\theta\,\dot\phi\left(-\operatorname{cosec}^2\theta\sin\phi\,\dot\theta + \cot\theta\cos\phi\,\dot\phi\right)$$
$$= 0\,.$$

For the third Killing vector, we calculate the Lie bracket of the second and third Killing vectors to get

$$\partial_\phi(0,0,-\cos\phi,\cot\theta\sin\phi) - (-\cos\phi\,\partial_\theta - \cot\theta\sin\phi\,\partial_\phi)(0,0,0,1)$$
$$= (0,0,\sin\phi,\cot\theta\cos\phi)\,.$$

8.2   The first statement is a consequence of $\partial L/\partial t = 0 = \partial L/\partial\phi$. For large r, the metric is that of Minkowski space, where $\dot t = \gamma(u) \geq 1$. Therefore we must have $E \geq 1$ for escape.

By substituting for $\dot t$ and $\dot\phi$ in the four-velocity condition

$$\left(1 - \frac{2m}{r}\right)\dot t^2 - \frac{\dot r^2}{1 - 2m/r} - r^2\dot\phi^2 = 1$$

(the orbit is equatorial, so $\theta = \pi/2$), we obtain

$$\dot r^2 + 1 + \frac{J^2}{r^2} - \frac{2m}{r} - \frac{2mJ^2}{r^3} = E^2\,.$$

Hence by differentiating with respect to r, and using $\frac{1}{2}\mathrm{d}(\dot r^2)/\mathrm{d}r = \ddot r$, we have

$$\ddot r - \frac{J^2}{r^3} + \frac{m}{r^2} + \frac{3mJ^2}{r^4} = 0\,, \tag{A.1}$$

and hence the given result.

For a circular orbit of radius R, we have

$$\dot r = 0 = \ddot r, \qquad r = R\,,$$

and hence

$$J^2\left(-\frac{3m}{R^4} + \frac{1}{R^3}\right) = \frac{m}{R^2}\,,$$

which gives $J^2 = mR^2/(R - 3m)$. Also

$$E^2 = \left(1 + \frac{J^2}{R^2}\right)\left(1 - \frac{2m}{R}\right) = \frac{(R - 2m)^2}{R(R - 3m)}\,.$$

Therefore

$$\dot t = \frac{RE}{R - 2m} = \frac{\sqrt{R}}{\sqrt{R - 3m}}\,.$$

It follows that

$$\left(\frac{\mathrm{d}\phi}{\mathrm{d}t}\right)^2 = \left(\frac{\dot\phi}{\dot t}\right)^2 = \frac{J^2}{R^4}\frac{R-3m}{R} = \frac{m}{R^3}\,.$$

By substituting $r = R + \epsilon$ into (A.1), and discarding second-order terms in ϵ and its derivatives, we find

$$\ddot\epsilon + \frac{m}{R^2} - \frac{2m\epsilon}{R^3} - \frac{J^2}{R^3} + \frac{3J^2\epsilon}{R^4} + \frac{3mJ^2}{R^4} - \frac{12mJ^2\epsilon}{R^5} = 0\,,$$

which gives

$$\ddot\epsilon + \frac{m(R-6m)\epsilon}{R^3(R-3m)} = 0\,,$$

and hence that the orbit is stable if and only if $R > 6m$ (note that $R < 3m$ is not possible).

8.3 For null (nonradial) equatorial geodesics, the geodesic equations reduce to

$$p^2 = \alpha^2 + 2u^3 - u^2\,,$$

where $u = m/r$ and $p = \mathrm{d}u/\mathrm{d}\phi$. By working from the given equation, we have

$$\log A + \phi = \log(1 - 3u) - 2\log\left(\sqrt{3} + \sqrt{1 + 6u}\right)\,.$$

Hence by differentiating both sides with respect to u, we have

$$
\begin{aligned}
p^{-1} &= -\frac{3}{1-3u} - \frac{6}{\sqrt{1+6u}\,(\sqrt{3}+\sqrt{1+6u})} \\[2mm]
&= -\frac{3}{1-3u}\left(1 + \frac{\sqrt{3}-\sqrt{1+6u}}{\sqrt{1+6u}}\right) \\[2mm]
&= -\frac{3}{1-3u}\frac{\sqrt{3}}{\sqrt{1+6u}}\,.
\end{aligned}
$$

Therefore

$$p^2 = \frac{(1-3u)^2(1+6u)}{27} = \frac{1}{27} - u^2 + 2u^3\,,$$

thus we have a solution of the geodesic equation. As $\phi \to -\infty$, for $A > 0$ the orbit spirals in from infinity, asymptotic to the null geodesic orbit at $r = 3m$; for $A < 0$, it spirals out towards it from the region $3m > r > 2m$.

11.2   You will find it helpful to establish

$$w_{ab}(0, \boldsymbol{r}) = \frac{1}{2} \int \left(k_{ab}(\boldsymbol{n}) + \overline{k}_{ab}(-\boldsymbol{n}) \right) \exp(\mathrm{i}\boldsymbol{n} \,.\, \boldsymbol{r}) \, \mathrm{d}V$$

and to use the Fourier inversion theorem

$$f(\boldsymbol{r}) = \int \hat{f}(\boldsymbol{n}) \exp(\mathrm{i}\boldsymbol{n} \,.\, \boldsymbol{r}) \, \mathrm{d}V',$$

$$\hat{f}(\boldsymbol{n}) = \frac{1}{(2\pi)^3} \int f(\boldsymbol{r}) \exp(-\mathrm{i}\boldsymbol{n} \,.\, \boldsymbol{r}) \, \mathrm{d}V,$$

where $\mathrm{d}V'$ is the volume element in the space of $\boldsymbol{n}$s.

11.5   Those who have studied exterior calculus will be able to deduce this directly from the closure of the three-form $J^a \varepsilon_{abcd} \, \mathrm{d}x^b \wedge \mathrm{d}x^c \wedge \mathrm{d}x^d$. An alternative direct method is to observe first that because Q is invariant, it is only necessary to show that $\partial Q / \partial t' = 0$. Write

$$Q = \int_V \left([\rho] - \boldsymbol{e} \,.\, [\boldsymbol{j}] \right) \mathrm{d}V,$$

where $\boldsymbol{e} = (\boldsymbol{r}' - \boldsymbol{r})/|\boldsymbol{r}' - \boldsymbol{r}|$ and ρ and $\boldsymbol{j}$ are the temporal and spatial parts of the four-current J^a. Now find the derivative with respect to t' by differentiating under the integral sign. The key steps are to use the divergence theorem, the continuity equation

$$\frac{\partial \rho}{\partial t} + \boldsymbol{\nabla} \,.\, \boldsymbol{j} = 0,$$

and the identity

$$\boldsymbol{\nabla}[f] = [\boldsymbol{\nabla}f] + [\partial_t f]\boldsymbol{e}, \qquad (\text{A.2})$$

which should be derived for a general function $f(t, \boldsymbol{r})$. Here $\boldsymbol{\nabla}$ is the gradient with respect to the spatial coordinates x, y, z.

12.1   To show existence, prove from the completeness of ω and from the fact that $\mathrm{d}t/\mathrm{d}\tau > 1$ that there is a value of τ for which $t = x^0(\tau) = 0$. Use the intermediate value theorem to deduce that there is an event on ω at which $t \le 0$ and $t^2 - x^2 - y^2 - z^2 = 0$. To show uniqueness, show that any two events on ω are connected by a timelike vector and then show that it is not possible to express a timelike vector as the difference between two future-pointing null vectors.

12.2   You need to show that $n^a \nabla_a n^b \propto n^b$ on Σ, which you can do by showing that $X_b n^a \nabla_a n^b = 0$ on Σ for every vector X^a such that $X^a n_a = 0$.

12.3 Suppose that g_{ab} and g_{ab} are metrics on a space–time. Write $\tilde{g}_{ab} \leq g_{ab}$ if every vector which is timelike with respect to $\tilde{g}_{ab}$ is also timelike with respect to g_{ab}. Prove that this is a partial ordering. Show that if $I^-(\omega)$ and $\tilde{I}^-(\omega)$ are the pasts of a worldline ω with respect to the two metrics, and that if $\tilde{g}_{ab} \leq g_{ab}$, then $\tilde{I}^-(\omega) \subseteq I^-(\omega)$. By considering the Kerr metric in the form (10.4) and by taking $\tilde{g}_{ab} = g_{ab} - kt_a t_b$ for some constant k, construct a flat space–time metric on $r > r_0$ with $\tilde{g}_{ab} \leq g_{ab}$, where g_{ab} is the Kerr metric.

12.7 It is helpful to start by writing $v^2 - w^2 = (v + w)(v - w)$, $\mathrm{d}v^2 - \mathrm{d}w^2 = (\mathrm{d}v + \mathrm{d}w)(\mathrm{d}v - \mathrm{d}w)$, and

$$\mathrm{d}r^2 + r^2\,\mathrm{d}\theta^2 + r^2\sin^2\,\mathrm{d}\varphi^2 = \mathrm{e}^{-2t}\left((\mathrm{d}x - x\,\mathrm{d}t)^2 + (\mathrm{d}y - y\,\mathrm{d}t)^2 + (\mathrm{d}z - z\,\mathrm{d}t)^2\right).$$

Note that $x^2 + y^2 + z^2 = r^2\mathrm{e}^{2t}$.

Appendix B: Further Problems

The problems that follow are taken from final examination papers set in Oxford over the past 15 years, in some cases adapted for notational consistency with the text, and with any hints deleted. The passage of time and the conventions of anonymity and collective responsibility of examiners make it hard to identify all the original authors; some may even be borrowed from other texts. I must therefore apologise for including them without acknowledgement. A few I recognise as my own, others are likely to be by my colleagues, Roger Penrose, Paul Tod, and Lionel Mason.

B.1   A model universe has metric

$$ds^2 = dt^2 - R(t)^2 \left(dr^2 + \sin^2 r (d\theta^2 + \sin^2 \theta d\varphi^2) \right),$$

where $R(t) > 0$ for $t \in (t_0, t_1)$. Obtain the geodesic equations and write down the Christoffel symbols (with $x^0 = t$, $x^1 = r$, $x^2 = \theta$, $x^3 = \varphi$). Show that there are geodesics on which θ and r are constant and equal to $\pi/2$. Show that if

$$\int_{t_0}^{t_1} \frac{dt}{R} < 2\pi,$$

then a photon cannot make a complete circuit of the circle $\theta = r = \pi/2$, $0 \le \varphi \le 2\pi$ between $t = t_0$ and $t = t_1$.

B.2   Let ∇ denote the Levi-Civita connection in a curved space–time. Show that there is a tensor $R_{abc}{}^d$ such that

$$(\nabla_a \nabla_b - \nabla_b \nabla_a) X^d = R_{abc}{}^d X^c$$

for every X^a. Show that

$$(\nabla_a \nabla_b - \nabla_b \nabla_a) T_{cd} = -R_{abc}{}^e T_{ed} - R_{abd}{}^e T_{ce}$$

for every T_{ab}. Show that if $F_{ab} = F_{[ab]}$ satisfies Maxwell's equations $\nabla_a F^{ab} = 0$ and $\nabla_{[a} F_{bc]} = 0$, then

$$\nabla_a \nabla^a F_{bc} + 2R_{abcd} F^{ad} - 2R_{[b}{}^d F_{c]d} = 0.$$

B.3   The Riemann curvature tensor $R_{abc}{}^d$ in a curved space–time endowed with a torsion-free connection satisfies the equation

$$(\nabla_a \nabla_b - \nabla_b \nabla_a) X^d = R_{abc}{}^d X^c$$

and can be expressed in the form

$$R_{abc}{}^d = \partial_a \Gamma^d{}_{bc} - \partial_b \Gamma^d{}_{ac} + \Gamma^d{}_{ae} \Gamma^e{}_{bc} - \Gamma^d{}_{be} \Gamma^e{}_{ac}.$$

By using this show that the curvature tensor has the following symmetries.

$$R_{abcd} = -R_{bacd}, \quad R_{abcd} = R_{cdab}, \quad R_{abcd} = -R_{abdc}, \quad R_{[abc]d} = 0.$$

Now let Φ_a be the electromagnetic four-potential, so that the electromagnetic field tensor is $F_{ab} = \partial_a \Phi_b - \partial_b \Phi_a$, and satisfies the free-space Maxwell's equations

$$\nabla_a F^{ab} = 0, \qquad \nabla_{[a} F_{bc]} = 0.$$

Show that the second of these equations is satisfied for any four-potential Φ_a, but that the first holds only if

$$\nabla_b \nabla^b \Phi_a - \nabla_a (\nabla^b \Phi_b) + R_{ab} \Phi^b = 0,$$

where R_{ab} is the Ricci tensor.

Suppose, in addition, that the Lorenz condition $\nabla_a \Phi^a = 0$ holds. Show that when the space–time scale of variations of S is much smaller than those of C and of the connection, and second derivatives of S may be ignored, there are approximate solutions of the form $\Phi^a = C^a \exp(iS)$ (the *geometrical optics* approximation), provided that $k_a = \nabla_a S$ is a null covector field. By considering $\nabla_a(k^b k_b)$ show that the integral curves of k are null geodesics.

B.4   Explain briefly how the geodesic hypothesis for free particles and photons can be justified from the principle that special relativity should hold over short times and distances in frames in free-fall.

A space–time has metric

$$ds^2 = dt^2 - dx^2 - dy^2 - dz^2 + 2\varphi(dt + dz)^2,$$

where φ is a function of x and y alone.

(a) Show that if $\big(t(\tau), x(\tau), y(\tau), z(\tau)\big)$ is a solution of the geodesic equation, then

$$\frac{\mathrm{d}x}{\mathrm{d}\tau} = \alpha\frac{\partial\varphi}{\partial x}, \qquad \frac{\mathrm{d}y}{\mathrm{d}\tau} = \alpha\frac{\partial\varphi}{\partial y},$$

for some constant α, where τ is proper time.

(b) Suppose that O and O' are observers with worldlines on which x, y, and z are constant. By considering an appropriate constant of the motion for the photon, show that if O sends a photon to O' and if the frequencies of the photon as measured by O and O' are ω and ω', then

$$\frac{\omega}{\omega'} = \sqrt{\frac{1 + 2\varphi(O')}{1 + 2\varphi(O)}}.$$

B.5   Let m_{ab} and m^{ab} be the covariant and contravariant metric tensors on Minkowski space, $\mathbb{M}$, with standard inertial coordinates x^a so that

$$(m_{ab}) = \begin{pmatrix} 1 & 0 & 0 & 0 \\ 0 & -1 & 0 & 0 \\ 0 & 0 & -1 & 0 \\ 0 & 0 & 0 & -1 \end{pmatrix} = (m^{ab}).$$

Let n_a be a constant null covector on $\mathbb{M}$, and define a new metric on $\mathbb{M}$ by

$$g_{ab} = m_{ab} + n_a n_b f,$$

where f is a function on $\mathbb{M}$ such that

$$m^{ab} n_a \partial_b f = 0,$$

and where $\partial_a = \partial/\partial x_a$. Show that the connection derived from g_{ab} is given by

$$\Gamma^a{}_{bc} = m^{da}\left[n_d n_{(b}\partial_{c)}f - \tfrac{1}{2}n_b n_c \partial_d f\right].$$

Show that the Ricci tensor is

$$R_{ab} = \frac{1}{2}n_a n_b \square f,$$

where $\square = m^{ab}\partial_a\partial_b$, and that Einstein's vacuum field equations in this case can have plane wave solutions provided the propagation vector k^a satisfies $n_a k^a = 0$. Deduce that the Ricci scalar vanishes and that then the Ricci tensor satisfies the conservation equation

$$\nabla^a R_{ab} = 0.$$

B.6   A space–time M has metric

$$ds^2 = dt^2 - \alpha(t)^2(dx^2 + dy^2 + dz^2)\,.$$

You may assume without calculation that the nonzero components of the Ricci tensor for M are given by

$$R_{tt} = 3\alpha''/\alpha; \qquad R_{xx} = R_{yy} = R_{zz} = -\alpha\alpha'' - 2\alpha'^2\,,$$

where $\alpha' = d\alpha/dt$.

The space–time M is filled with dust of rest mass density ρ whose four-velocity is orthogonal to the surfaces of constant t. Show that Einstein's field equations reduce to the two equations

$$\begin{aligned}
3\alpha'' &= -4\pi G\rho\alpha \\
\alpha''\alpha + 2(\alpha')^2 &= 4\pi G\rho\alpha^2\,.
\end{aligned}$$

By assuming that α and α' are both nonnegative, deduce that the general solution is

$$\alpha(t) = A(t - t_0)^{2/3}\,,$$

where A and t_0 are constants.

B.7   A null geodesic γ lies in the equatorial plane $\theta = \pi/2$ of the Schwarzschild metric, which in conventional coordinates is given by:

$$ds^2 = (1 - 2m/r)dt^2 - (1 - 2m/r)^{-1}dr^2 - r^2(d\theta^2 + \sin^2\theta d\varphi^2)\,.$$

Write down the geodesic equations, and hence show that along γ,

$$p^2 = 2u^3 - u^2 + \alpha^2\,,$$

where $u = m/r$, $p = du/d\varphi$, and α is a constant. Sketch the trajectories in the (u, p) phase-plane, both in the region $0 < u < 1/2$ and also in the region $u > 1/2$.

For the case $\alpha = 0$ and $u > 1/2$, show that the geodesic has an equation of form

$$r = 2m\cos^2(1/2(\varphi - \varphi_0)); \qquad \theta = \pi/2; \qquad t = t_0$$

where φ_0 and t_0 are constants. Indicate by a sketch where this geodesic is located in the complete Kruskal extension of the Schwarzschild solution.

B.8   A space–time metric has the form

$$ds^2 = f(r)^2 d\tau^2 - dr^2 - dy^2 - dz^2 \,,$$

where f is a positive function of r. Two nearby observers A and B have respective worldlines given by

(A) $y = z = 0, \quad r = r_0,$ (B) $y = z = 0, \quad r = r_1,$

where $r_0 < r_1$ are constants. Show that τ is a constant multiple of proper time on each worldline. Are the worldlines geodesic? Give reasons for your answer.

A light signal emitted by A at $\tau = \tau_0$ is received by B at proper time $\tau = \tau_1$ and immediately reflected back to A, where it arrives at $\tau = \tau_2$. Show that

$$\tau_1 - \tau_0 = \int_{r_0}^{r_1} \frac{dr}{f(r)} \,.$$

Deduce that light emitted by A with frequency ω is seen by B to have frequency $\omega f(r_0)/f(r_1)$. Deduce also that

$$\tau_2 - \tau_0 = 2 \int_{r_0}^{r_1} \frac{dr}{f(r)} \,,$$

and hence that if A measures the distance to B by the radar method, then this distance is constant.

Show that when $f = r$, the metric can be reduced to the Minkowski metric by a coordinate change, and that the worldlines become

$$t = r \sinh \tau, \qquad x = r \cosh \tau, \qquad y = z = 0,$$

for two constant values of r.

Explain why the observed redshift of light travelling from the bottom to the top of a tower in the earth's gravitational field is incompatible with any special relativistic theory of gravity in which photon worldlines are null geodesics and the frame of the tower is inertial.

When $f = r$, the worldlines of A and B are a constant distance apart (as measured by both observers), but there is a redshift for light travelling from one to the other. Explain why this does not contradict your answer.

B.9   (i) You are given that Maxwell's equations in curved space–time are

$$\nabla_{[a} F_{bc]} = 0, \qquad \nabla^a F_{ab} = 0 \,,$$

where $F_{(ab)} = 0$ and where ∇ is the Levi-Civita connection. Show that if $F_{ab} = \nabla_{[a}\Phi_{b]}$ for some covector field Φ such that $\nabla_a \Phi^a = 0$, then the first equation is satisfied identically, and the second reduces to

$$\nabla_b \nabla^b \Phi_a = -R_{ab}\Phi^b.$$

(ii) Let f_{ab} be a nonzero skew-symmetric tensor at an event and suppose that $f^{ab}\alpha_b = 0$ and $f_{[ab}\alpha_{c]} = 0$ for some nonzero covector α_a. Show that α_a is null.

(iii) Suppose that u is a smooth function on space–time such that its gradient $\alpha_a = \nabla_a u$ is nonzero on the hypersurface S defined by $u = 0$; and suppose that F_{ab} is a solution of Maxwell's equations with the property that the tensor $f_{ab} = u^{-1}F_{ab}$ is smooth on S. Show that $F_{ab} = 0$ on S. Show also that if f_{ab} is nonzero on S, then S is null (i.e., α_a is null on S).

Comment on the physical significance of the fact that S must be null.

B.10   A spherically symmetric space–time metric has the form

$$ds^2 = A(r)\,dt^2 - \frac{dr^2}{A(r)} - r^2\,d\theta^2 - r^2 \sin^2\theta\,d\varphi^2 \quad (r > 0).$$

Write down the geodesic equations and show that there are null nonradial geodesics on which θ takes the constant value $\pi/2$. Show that such geodesics are given by
$$p^2 = k - u^2 Q(u),$$
where $u = 1/r$, $p = du/d\varphi$, $Q(u) = A(1/u)$, and k is a positive constant (depending on the geodesic). Hence show that these nonradial null geodesics are given by

$$\frac{d^2 u}{d\varphi^2} = -uQ(u) - \tfrac{1}{2}u^2 Q'(u).$$

(i) Show that if $A(r) = 1 - 2m/r$, then there is a photon orbit on which r takes the constant value $3m$.

(ii) Suppose that $A(r) = Q(u)$, where Q is a polynomial in u such that $Q > 0$ for all $u > 0$. Show that there are photon orbits on which r takes any one of the constant values $r = 1/u_i$, where $0 < u_1 \le u_2 \le \cdots$ are the positive roots of $u^2 Q'(u) + 2uQ(u)$. Suppose that the roots of this polynomial are distinct. Is the orbit at $r = 1/u_2$ stable?

B.11 Define the *Riemann tensor R_{abcd}* of a space–time metric and write down its symmetries.

Show that for any vector fields X, Y,

$$[X, Y]^b = X^a \nabla_a Y^b - Y^a \nabla_a X^b \,,$$

where $[X, Y]$ has components $[X, Y]^b = X^a \partial_a Y^b - Y^a \partial_a X^b$ in some coordinate system x^a, ∇ is the Levi-Civita connection, and ∂_a denotes $\partial/\partial x^a$.

Show that if $X^a V_a$ is constant along every affinely parametrized geodesic $x^a = x^a(s)$, where $V^a = \mathrm{d}x^a/\mathrm{d}s$, then X satisfies the *Killing equation*:

$$\nabla_a X_b + \nabla_b X_a = 0 \,.$$

Deduce that

$$\nabla_a \nabla_c X_b = R_{cbad} X^d \,,$$

and hence that X^a satisfies the *Jacobi equation*

$$D^2 X^d = R_{abc}{}^d V^a X^b V^c, \qquad (D = V^a \nabla_a)$$

along any geodesic.

Let X, Y be solutions to the Killing equation. Show that if $X^a = Y^a$ and $\nabla_a X^b = \nabla_a Y^b$ at some event P, then $X^a = Y^a$ along every geodesic through P. (You must state clearly any theorems that you use about the uniqueness of solutions of systems of second-order ordinary differential equations.)

Deduce that the space of solutions to the Killing equation has dimension at most 10. Give an example of a space–time in which the dimension is equal to 10.

B.12 Let g_{ab} be a general space–time metric. Show that for any event A, there exists a coordinate system x^a such that $\partial_a g_{bc} = 0$ at A, where $\partial_a = \partial/\partial x^a$.

Show that in such a coordinate system,

$$\Gamma^a_{bc} = 0 \quad \text{and} \quad R_{abcd} = \tfrac{1}{2}[\partial_a \partial_c g_{bd} + \partial_b \partial_d g_{ac} - \partial_a \partial_d g_{bc} - \partial_b \partial_c g_{ad}]$$

at the event A.

Show that there does not exist a coordinate transformation that reduces the metric

$$\mathrm{d}s^2 = (1 + x^2)\mathrm{d}t^2 - \mathrm{d}x^2 - \mathrm{d}y^2 - \mathrm{d}z^2$$

to the metric of Minkowski space.

B.13 The gravitational field of a spherically symmetric black hole is represented by the Schwarzschild metric

$$ds^2 = \left(1 - 2m/r\right)dt^2 - \left(1 - 2m/r\right)^{-1}dr^2 - r^2 d\theta^2 - r^2 \sin^2\theta d\varphi^2.$$

Explain briefly the sense in which $r = 2m$ is only an apparent singularity.

A particle in free-fall has worldline $(t, r, \theta, \varphi) = (t(\tau), r(\tau), \pi/2, \varphi(\tau))$, where τ is proper time and $r > 2m$. Show that

$$E = \left(1 - \frac{2m}{r}\right)\dot{t} \quad \text{and} \quad J = r^2\dot{\varphi}$$

are constant along the worldline, where the dot denotes differentiation with respect to τ. Explain why the particle cannot escape to infinity if $E < 1$.

Show that

$$\frac{E^2 - \dot{r}^2}{1 - 2m/r} - \frac{J^2}{r^2} = 1.$$

Deduce that if $E = 1$ and $J = 4m$, then

$$\frac{\sqrt{r} - 2\sqrt{m}}{\sqrt{r} + 2\sqrt{m}} = A e^{\epsilon\varphi/\sqrt{2}},$$

where $\epsilon = \pm 1$ and A is a constant. Describe the orbit that starts at $\varphi = 0$ in each of the cases (i) $A = 0$, (ii) $A = 1$, $\epsilon = -1$, (iii) $A = (\sqrt{3} - 2)/(\sqrt{3} + 2)$, $\epsilon = -1$.

B.14 A space–time has the metric

$$ds^2 = g_{ab}dx^a dx^b.$$

Show that the Christoffel symbol, defined by

$$\Gamma^a{}_{bc} = \frac{1}{2}g^{da}(\partial_c g_{bd} + \partial_b g_{cd} - \partial_d g_{bc}),$$

can be derived from this by using Lagrange's equations. Calculate all nonvanishing Christoffel symbols for the metric

$$ds^2 = 2du\, dv - A(u)dx^2 - B(u)dy^2.$$

Obtain the equations of the geodesics and show that, for each geodesic, we can find constants $\alpha, \beta, \gamma, \delta$ such that

$$v = \frac{1}{2}\int\left(\frac{\alpha^2}{A(u)} + \frac{\beta^2}{B(u)}\right)du + \gamma u + \delta.$$

B.15   The portion of space–time outside a black hole of mass m has the Schwarzschild metric

$$\mathrm{d}s^2 = (1 - 2m/r)\,\mathrm{d}t^2 - (1 - 2m/r)^{-1}\,\mathrm{d}r^2 - r^2\mathrm{d}\theta^2 - r^2\sin^2\theta\,\mathrm{d}\varphi^2.$$

Obtain the equations of the photon orbits. Show that a light ray passing the black hole at a distance $D \gg m$ is deflected through an angle of approximately $4m/D$.

Sketch the phase portrait of the equatorial null geodesics in the u, p plane, where $u = m/r < \frac{1}{2}$ and $p = \mathrm{d}u/\mathrm{d}\varphi$. A photon is emitted at $r = 3m + \epsilon$ in the equatorial plane in a direction orthogonal to the radius vector, where $|\epsilon| \ll m$. Describe the photon's orbit in the two cases $\epsilon > 0$ and $\epsilon < 0$, identifying the corresponding curves in the u, p-plane.

B.16   Let T be a four-vector field on a space–time with metric $\mathrm{d}s^2 =_{ab} \mathrm{d}x^a\mathrm{d}x^b$. Show that if the components T^a are constant in the coordinate system x^a, then

$$\nabla_a T_b = \partial_{[a}T_{b]} + \tfrac{1}{2}T^c\partial_c g_{ab},$$

where ∇ is the Levi-Civita connection. Deduce that if $T^c\partial_c g_{ab} = 0$, then

$$\nabla_{(a}T_{b)} = 0, \qquad \nabla_c\nabla_a T_b = R_{abcd}T^d,$$

and

$$R_{ab}T^aT^b = \tfrac{1}{2}\Box(T^cT_c) - (\nabla_a T_b)(\nabla^a T^b),$$

where $\Box = \nabla_a\nabla^a$.

Suppose that

$$\mathrm{d}s^2 = A(\mathrm{d}x^0)^2 - h_{\alpha\beta}\mathrm{d}x^\alpha\mathrm{d}x^\beta,$$

where $\alpha, \beta = 1, 2, 3$, with summation convention. Show that if A and $h_{\alpha\beta}$ are independent of x^0, then

$$R_{00} = -h^{-1/2}\partial_\alpha\left(h^{1/2}h^{\alpha\beta}\partial_\beta A\right) - \tfrac{1}{2}A^{-1}h^{\alpha\beta}(\partial_\alpha A)(\partial_\beta A),$$

where $h^{\alpha\beta}h_{\beta\gamma} = \delta^\alpha_\gamma$ and $h = \det h_{\alpha\beta}$.

B.17   A space–time has metric

$$g_{ab} = m_{ab} + \epsilon h_{ab}\,,$$

where m_{ab} is the Minkowski space metric, ϵ is a small parameter, and h_{ab} is symmetric, with $\partial_0 h_{ab} = O(\epsilon)$. A particle is in free-fall, with four-velocity $V = (1, \mathbf{0}) + O(\epsilon)$. Show that, if terms of order ϵ^2 are ignored, then its equation of motion is

$$\ddot{\mathbf{r}} = -\tfrac{1}{2}\nabla(\epsilon h_{00}) + O(\epsilon^2)\,.$$

How can this result be used to recover Newton's theory of gravity from general relativity for slow-moving bodies in the weak-field limit?

Describe the corresponding Newtonian gravitational field when the metric is

$$ds^2 = (1 + 2\epsilon z)dt^2 - dx^2 - dy^2 - dz^2\,.$$

Show that, if terms of order ϵ^2 are ignored, then the coordinate transformation

$$\hat{t} = (1 + \epsilon z)t, \qquad \hat{x} = x, \qquad \hat{y} = y, \qquad \hat{z} = z + \tfrac{1}{2}\epsilon t^2$$

reduces the metric to the Minkowski form. Explain this result in terms of the equivalence principle.

B.18   Show that the quantities

$$E = \left(1 - \frac{2m}{r}\right)\dot{t} \quad \text{and} \quad J = r^2 \sin^2\theta\,\dot{\varphi}$$

are constant along the timelike geodesics of the Schwarzschild metric

$$ds^2 = (1 - 2m/r)\,dt^2 - \frac{dr^2}{1 - 2m/r} - r^2(d\theta^2 + \sin^2\theta\,d\varphi^2)$$

(the dot denotes differentiation with respect to proper time).

A *stationary* observer is one on whose worldline r, θ, and φ are constant. Explain why there are no stationary observers at $r < 2m$. Show that if a particle in free-fall has speed v relative to a stationary observer at an event on its worldline, then

$$E = \frac{\sqrt{1 - 2m/r}}{\sqrt{1 - v^2}}\,.$$

Explain how the constancy of E reduces to a conservation law in Newtonian gravity (which you should identify) when m and v are small. What can you say about orbits on which $E < 1$?

Show that the nonradial equatorial timelike geodesics ($\theta = \pi/2$) are given by

$$\left(\frac{du}{d\varphi}\right)^2 = 2\beta^2 u + 2k - u^2 + 2u^3\,,$$

where $u = m/r$, $\beta = m/J$, and $k = (E^2 - 1)m^2/2J^2$. By making the substitution $u = (1/4)\cosh^2 x$ show that, if $k = 0$ and $\beta = 1/4$, then

$$\frac{\mathrm{d}x}{\mathrm{d}\varphi} = \frac{\pm \sinh x}{2\sqrt{2}}\,.$$

Verify that a possible solution is

$$\exp\left(\frac{\varphi}{2\sqrt{2}}\right) = \frac{1 + \mathrm{e}^x}{1 - \mathrm{e}^x}\,.$$

Describe the behaviour of the corresponding orbit as $\varphi \to \infty$.

B.19   In a space–time M, a vector field has components X^a and the metric has components g_{ab}, with respect to a coordinate system (x^0, x^1, x^2, x^3). In these coordinates, $X^a = (1, 0, 0, 0)$ and $\partial g_{ab}/\partial x^0$ is zero. Show that X^a satisfies the Killing equation

$$\nabla_a X_b + \nabla_b X_a = 0\,.$$

By differentiating this equation, deduce that X_a also satisfies the equation

$$\nabla_a \nabla_b X_c = R_{bcad} X^d\,,$$

where R_{abcd} is the Riemann tensor of M.

What is the geodesic deviation equation? Show that X^a satisfies the geodesic deviation equation along any geodesic in M.

Show that, if X^a and Y^a both satisfy the Killing equation, then so does their commutator Z^a, defined by

$$Z^a = X^b \nabla_b Y^a - Y^b \nabla_b X^a\,.$$

B.20   The Minkowski metric in inertial coordinates $(x^0, x^1, x^2, x^3) = (t, x, y, z)$ is

$$ds^2 = \mathrm{d}t^2 - \mathrm{d}x^2 - \mathrm{d}y^2 - \mathrm{d}z^2$$

and the wave operator is defined by

$$\Box u = \frac{\partial^2 u}{\partial t^2} - \frac{\partial^2 u}{\partial x^2} - \frac{\partial^2 u}{\partial y^2} - \frac{\partial^2 u}{\partial z^2}\,.$$

Explain why the wave operator in arbitrary coordinates $\tilde{x}^a$ can be written as

$$\Box u = g^{ab} \nabla_a \nabla_b u,$$

where you should explain what is meant by ∇_a.

Rindler coordinates (T, X, Y, Z) are given implicitly in terms of inertial coordinates (t, x, y, z) by

$$t = X \sinh T, \qquad x = X \cosh T, \qquad y = Y, \qquad z = Z.$$

Show that in these coordinates the metric becomes

$$ds^2 = X^2 \, dT^2 - dX^2 - dY^2 - dZ^2.$$

By using the geodesic equation, or otherwise, obtain the Christoffel symbols Γ^c_{ab} for this metric and show that

$$g^{ab} \Gamma^c_{ab} = X^{-1} \delta^c_1.$$

By using the formula obtained above for $\Box u$ obtain the wave equation in Rindler coordinates. Show that $u = f(Xe^T)$ is a solution for any smooth f.

B.21   In a space–time M, a timelike geodesic γ has four-velocity vector V^a. The vector-field Y^a defined along γ is a connecting vector to an infinitesimally neighbouring geodesic. Assuming the equations

$$DV^a = 0, \qquad DY^a = Y^b \nabla_b V^a,$$

where D is $V^b \nabla_b$, derive the geodesic deviation equation

$$D^2 Y^a = R_{bcd}{}^a V^b Y^c V^d.$$

Suppose that $Y^a V_a = 0$ at one point of γ. Deduce that $Y^a V_a = 0$ at all points of γ.

Now suppose that the Riemann tensor R_{abcd} of M can be written in terms of the metric g_{ab} and a function F in the form

$$R_{abcd} = F(g_{ac} g_{bd} - g_{ad} g_{bc}).$$

What are the Ricci tensor R_{ab} and the Ricci scalar R in terms of F and g_{ab}?

What is the contracted Bianchi identity and what can you deduce about F from it?

Show that with these assumptions the geodesic deviation equation in M becomes

$$D^2 Y^a = F Y^a.$$

Solve this equation by writing Y^a as $f X^a$ where X^a is parallelly propagated along γ and f is a function to be found. Show that if $F < 0$ then Y^a necessarily has a zero in any piece of γ with proper length greater than $2\pi/\sqrt{-F}$.

B.22   The Schwarzschild metric is given in coordinates $(x^u) = (t, r, \theta, \varphi)$ by

$$ds^2 = (1 - 2m/r)\, dt^2 - (1 - 2m/r)^{-1}\, dr^2 - r^2(d\theta^2 + \sin^2\theta\, d\varphi^2)\,.$$

How may the geodesic equations for this metric be obtained from Lagrange's equations?

Show that there are geodesics confined to the equatorial plane $\theta = \pi/2$ and that these geodesics are determined by the equations

$$
\begin{aligned}
(1 - 2m/r)\,\dot{t} &= E\,, \\
r^2\dot{\varphi} &= J\,, \\
\dot{r}^2 &= E^2 - \left(1 - \frac{2m}{r}\right)\left(\mu + J^2/r^2\right)\,,
\end{aligned}
$$

where J, E, and μ are constants and the dot denotes $d/d\tau$. What is the significance of μ?

Hence or otherwise, deduce the equation

$$\ddot{r} = -\frac{1}{r^4}\left(m\mu r^2 - J^2 r + 3mJ^2\right)\,.$$

Show that for each J with $J^2 > 12m^2$ there are two timelike circular orbits at constant values of r, and for $J > 0$ there is a unique null circular orbit at a value r_0 of r, which you should find.

By setting $r = r_0 + \zeta$ for small ζ, or otherwise, determine whether the circular null orbit is stable.

B.23   Let ∇_a denote the Levi-Civita connection in a curved space–time. Write down a formula for $\nabla_a\nabla_b V^d - \nabla_b\nabla_a V^d$, where V^a is a vector field, in terms of V^a and the Riemann tensor $R_{abc}{}^d$.

Assuming the existence of local inertial coordinates, show that

$$
\begin{aligned}
R_{abcd} &= R_{[ab][cd]} \\
R_{[abc]d} &= 0 \\
R_{abcd} &= R_{cdab}\,.
\end{aligned}
$$

Show that for any covariant tensor field T_{ab},

$$\nabla_a\nabla_b T_{cd} - \nabla_b\nabla_a T_{cd} = -R_{abc}{}^e T_{ed} - R_{abd}{}^e T_{ce}\,,$$

where ∇_a is the Levi-Civita connection.

Show that if $R_{ab} = 0$ and that if F_{ab} is a skew-symmetric tensor such that $\nabla_{[a}F_{bc]} = 0$ and $\nabla^a F_{ab} = 0$, then

$$\Box F_{ab} = R_{abcd}F^{cd}\,,$$

where $\Box = \nabla_a \nabla^a$.

B.24   The Schwarzschild metric is

$$ds^2 = (1 - 2m/r)\,dt^2 - \frac{dr^2}{1 - 2m/r} - r^2(d\theta^2 + \sin^2\theta\,d\varphi^2)\,.$$

Show that the coordinate transformation

$$v = t + r + 2m\log(r - 2m)\,,$$

changes it to the form

$$ds^2 = (1 - 2m/r)\,dv^2 - 2dv\,dr - r^2(d\theta^2 + \sin^2\theta\,d\varphi^2)\,.$$

Show that the radial null geodesics are given by

$$v = \text{constant}\qquad\text{or}\qquad \left(1 - \frac{2m}{r}\right)\frac{dv}{dr} - 2 = 0\,.$$

Explain how this metric models the interior and exterior of a black hole. How would you show that the singularity at $r = 0$ is not an artefact of the choice of coordinates?

Show that

$$E = \left(1 - \frac{2m}{r}\right)\dot{v} - \dot{r}$$

is constant along timelike geodesics, where the dot denotes differentiation with respect to proper time.

Show that along the worldline of a particle falling radially into the black hole with $E = 1$,

$$\dot{r} = -\sqrt{\frac{2m}{r}}\,.$$

Show that the particle reaches the singularity at $r = 0$ in finite proper time.

B.25   What does it mean for a connection to be *torsion-free*? Show that the connection ∇_a defined by

$$\nabla_a V^b = \partial_a V^b + \tfrac{1}{2}g^{bd}V^c(\partial_a g_{cd} + \partial_c g_{ad} - \partial_d g_{ac})$$

for any vector field V^a is torsion-free and satisfies $\nabla_a g_{bc} = 0$. Show that these conditions uniquely determine the connection.

Show that

$$\nabla_a V^a = \frac{1}{\sqrt{|g|}} \frac{\partial}{\partial x^a} \left(\sqrt{|g|} V^a \right),$$

where g is the determinant of the matrix (g_{ab}).

For a general set of metric coefficients g_{ab} in a coordinate system x^a, write down the components of four independent solutions to the equation $\nabla_a V^a = 0$.

B.26   The Kerr metric in Boyer–Lindquist coordinates t, r, θ, ϕ is

$$ds^2 = \left(1 - \frac{2mr}{\Sigma} \right) dt^2 + \frac{4mar\sin^2\theta}{\Sigma} \, dt d\phi - \frac{\Sigma}{\Delta} dr^2 - $$

$$\Sigma \, d\theta^2 - \left(r^2 + a^2 + \frac{2ma^2 r \sin^2\theta}{\Sigma} \right) \sin^2\theta \, d\phi^2,$$

where $m > a > 0$ are constant parameters, $\Delta = r^2 + a^2 - 2mr$, and $\Sigma = r^2 + a^2 \cos^2\theta$.

Find two Killing vectors and explain how one of them, K^a, can be chosen so as to be timelike as $r \to \infty$ and can be used to define a notion of conserved energy for a freely falling test particle. By evaluating the energy to leading order as $r \to \infty$, justify the interpretation of the parameter m as the mass of the black hole. What is the interpretation of the parameter a? Explain briefly how this interpretation can be justified.

Find the values of r, r_+, and r_- with $r_+ > r_-$, on which the surfaces of constant r are null. What does it mean for $r = r_+$ to be an *event horizon*?

On which surfaces $r = f_\pm(\theta)$ does the vector field K^a become null? Explain why a particle with $r_+ < r < f_+(\theta)$ cannot remain at rest as viewed from infinity in the given coordinates. Draw a diagram in a plane containing the symmetry axis at constant t showing the location of $r = r_\pm$, $r = f_\pm$, and the singularities.

Bibliography

[1] R. Baum and W. Sheehan. *In Search of Planet Vulcan: The Ghost in Newton's Clockwork Universe*. Basic, New York (2003).

[2] L. Bod, E. Fischbach, G. Marx, and M. Náry-Ziegler. One hundred years of the Eötvös experiment. *Acta Physica Hungarica* **69**, 335–355 (1991).

[3] H. Bondi. *Assumption and Myth in Physical Theory*. Cambridge University Press, Cambridge, 1967.

[4] I. B. Cohen. Einstein's last interview. In: A. Robinson, ed., *Einstein, A Hundred Years of Relativity*. Palazzo, Bath, 2005.

[5] P. Coles. Einstein, Eddington, and the 1919 eclipse. In: Proceedings of International School on the Historical Development of Modern Cosmology, Valencia 2000, V. J. Martinez, V. Trimble, and M. J. Pons-Borderia, eds. ASP Conference Series, San Francisco (2001).

[6] *Gravity Probe B*. Web site http://einstein.stanford.edu/.

[7] A. Guth. *The Inflationary Universe*. Jonathan Cape, London (1997).

[8] S. W. Hawking and G. F. R. Ellis. *The Large-Scale Structure of Space–Time*. Cambridge University Press, Cambridge (1973).

[9] L. P. Hughston and K. P. Tod. *An Introduction to General Relativity*. London Mathematical Society Student Texts **5**. Cambridge University Press, Cambridge (1990).

[10] D. W. Jordan and P. Smith. *Nonlinear Ordinary Differential Equations : An Introduction to Dynamical Systems*. Third edition. Oxford University Press, Oxford (1999).

[11] R. P. Kerr. Gravitational field of a spinning mass as an example of algebraically special metrics. *Physical Review Letters*, **11**, 237–238 (1963).

[12] C. Lämmerzahl and G. Neugebauer. The Lens–Thirring effect: From the basic notions to the observed effects. *Lecture Notes in Physics*, **562**, 31–51 (2001).

[13] *LIGO (Laser Interferometer Gravitational Wave Observatory)*. Web site http://www.ligo.caltech.edu/.

[14] C. W. Misner, K. S. Thorne, and J. A. Wheeler. *Gravitation*. Freeman, San Francisco (1973).

[15] K. Nordtvedt. On the 'geodetic' precession of the lunar orbit. *Classical and Quantum Gravity* **13**, 1317–21 (1996).

[16] R. V. Pound and G. A. Rebka Jr. Apparent weight of photons. *Physical Review Letters*, **4**, 337 (1960).

[17] R. Penrose and R. M. Floyd. Extraction of rotational energy from a black hole. *Nature*, **229**, 177–9 (1971).

[18] R. Penrose. *The Road to Reality*. Jonathan Cape, London (2004).

[19] R. Penrose. *Techniques of Differential Topology in Relativity*. SIAM, Philadelphia, (1972).

[20] STEP. Web site http://einstein.stanford.edu/STEP/.

[21] J. H. Taylor, L. A. Fowler, and J. M. Weisberg. Measurements of general relativistic effects in the binary pulsar PSR1913+16. *Nature*, **277**, 437 (1979).

[22] R. W. Wald. *General Relativity*. University of Chicago Press, Chicago, 1984.

[23] N. M. J. Woodhouse. *Special Relativity*. Springer Undergraduate Mathematics Series, Springer, London (2003).